AF545022

Biotechnology in
Animal Health and Production

nipa
Browse Subject
eBooks / eChapters / Articles
eBooks
Explore Now
Browse, Search, Read & Buy...
Subject
Catalogues
eChapters
Publishers
Print Books
Forthcoming
eBooks
New Titles
scan for catalogue
NIPA
GENX
ONLINE RESOURCES
Publishing Books and Journals
Ebooks and e articles, Current Affairs
Reasoning, Logic and Aptitude
Competitive Examination Preparation
Language Learning and Development Programme
Document Quality Checker and Improvement Tool
Effective Public Speaking, Presentation and Interpersonal Skills
Online Programmes for Professional Development
Personality Development and Human Values
PAY USING
UPI
PayPal

Biotechnology in Animal Health and Production

Dr. S.K.Jindal, M.Sc., Ph.D.
Principal Scientist
Central Institute for Research on Goats
Makhdoom, Farah, Mathura, U.P.

Prof (Dr.) M.C.Sharma, M.V.Sc., Ph.D.
FNAAS, FISVM, FAAVS
Director
Indian Veterinary Research Institute
Izatnagar, U.P.

NEW INDIA PUBLISHING AGENCY
New Delhi-110 034

Published by
Sumit Pal Jain *for*
New India Publishing Agency
101, Vikas Surya Plaza, CU Block, L.S.C. Mkt.,
Pitam Pura, New Delhi- 110 088, (India)
Phone: 011-27341717, Fax: 011-27341616
Mobile : 09717133558
E-mail: newindiapublishingagency@gmail.com
Web: www.bookfactoryindia.com

ISBN : 978-93-80235-35-6

Comosed and Designed by NIPA

Preface

Biotechnology has brought revolution in our lives and is our future tool to reshape destiny. It has touched almost every aspect of human endeavour. Biotechnology is all set to have a massive impact on almost all aspects of our life in the 21st century. Animal Production and Health is an important area affecting human welfare especially the livestock farmers. Biotechnology has made inroads in this area and changed the way the animal production and health aspects are handled and holds great promises for future. India is fast pacing in the area of biotechnology to harness all potentials of productivity, growth and sustainable development. Animal biotechnology is the application of scientific processing or production of materials by animals to provide goods and services. Animal genetics is important in the application of Biotechnology to manage genetic disorders and improve animal breeding. Genomics, proteomics and bioinformatics are also being applied to animal biotechnology. This small book is an attempt to unravel the mysteries of biotechnology as it affect animal health and production. Most of the aspects in the related areas are discussed briefly and a small number of important references given as suggested reading for readers who wish to persue further. Since this is a new and important area, it is hoped that this book will stimulate minds to understand the biotechnological advances and will be useful to students of agriculture, animal husbandry and veterinary sciences. I am sure this topic will be covered as a separate subject in the coming years as part of the agriculture and veterinary curriculum and this book shall become an essential reading for such courses.

The authors are grateful to a number of individuals for their encouragement, guidance and help in preparation of this manuscript. Although it is difficult of list all of them but a few would definitely be mentioned. We are grateful to Dr. M.K. Tripathi, Dr. P.K. Rout, Dr. Ravi Ranjan, Dr. V.K. Bharti, Ms. Sonia Saraswat, Ms. Sushma Yadav who were helpful in reading the initial drafts and making a number of useful suggestions. The authors would like to express gratitude to numerous scientists of the world whose publications were consulted by the authors in preparation of this manuscripts. Last but not the least the authors are grateful to their respective spouses (Mrs. Shalini Sharma and Mrs. Sadhana Jindal) for their understanding, cooperation and patience. A special thanks to the publishers who took utmost interest in publication of this book.

Although this is an initial attempt, it is hoped that readers will bring to the notice of the authors any shortcomings, mistakes and suggestions for future improvement. Objective, criticism and suggestions will help us to improve our efforts. Happy reading.

S.K. JINDAL
M.C. SHARMA

Contents

Preface v

Chapter-1: Introduction to Biotechnology **1**

Biotechnology Products and Technologies, 5
Bottlenecks in Fast Progress of Animal Biotechnology, 6
Why Biotechnology should be Used in Animal Husbandry and Dairy Production, 7
Contribution of Livestock Sector to Food Basket, 7
Biotechnology and Environment, 10
Biotechnology of Nutrient Partitioning, 11
Genetic Programming for Better Animals, 13
Biotechnology of Reproduction, 14
Biotechnology of Animal Health, 14

Chapter-2: Biotechnology in Animal Production **17**

Reproductive Biotechnology, 18
Artificial Insemination (AI), 18
History and Future Perspectives of Artificial Insemination, 18
Semen Cryopreservation, 20
Buffalo, 23
Goat, 23
Cryoprotection and Anti Oxidants, 26
Cryopreservation and Genetic Stability of Livestock Spermatozoa, 26
Insemination Technique, 29
Cattle and Buffalo, 29
Sheep and Goat, 29

Conception Rate with Artificial Insemination, 30
Multiple Ovulation and Embryo Transfer (MOET), 31
Advantages of Embryo Transfer, 31
MOET Includes the Following Steps, 32
Superovulatory Treatments, 32
Insemination, 33
Recovery of Embryos, 33
Synchronization of Recipients, 34
Cattle, 34
Buffaloes, 35
Goats, 35
Horses, 36
In vitro Maturation, fertilization and culture of Oocytes, 36
Historical Development, 36
In Vitro Maturation of Oocytes, 39
In vitro Sperm Capacitation, 40
Scope, 43
Limitations, 44
Embryo Cryopreservation, 44
Ultrasonography in Reproductive Management, 47
Ultrasound-guided Transvaginal Oocyte Pick-up (OPU), 51
Embryo Cloning or Nuclear Transfer, 53
Blastomere Separation, 53
Embryo Splitting, 53
Nuclear Transfer Cloning (NT), 54
Embryonic Stem Cell Complementations, 54
Limitations and Future Prospects of Cloning, 56
Sexing of Embryos and Separation of X and Y Chromosomes Bearing Sperms, 57
PCR in Sex Determination, 58
Transgenic Animal Production, 60
Gene Transfer Through Direct DNA Microinjection, 65
Gene Transfer with Retroviral-Vector Assistance, 66
Gene Transfer through Production of Germ-line Chimeras, 67
Parthenogenetic Embryos: An Emerging Tool in Reproductive Biotechnology for Multiplication of Superior Germ Plasm, 68
Applications of Parthenogenesis, 71
Livestock Genomics and Marker Assisted Selection (Use of DNA Level Markers), 71
Stem Cells, 72
Vaccine for Fertility Control and Increase Prolificacy, 76
Auto-immunization to Increase Prolificacy, 77
Biotechnology of Conservation of Livestock Breed Biodiversity, 77
Gene for Prolificity: Booroola Merino, 78
Role of Biotechnology in Enhancing Production, 79

Milk, 80
Biotechnology of Rumen Manipulation, 82
Genetically Modified Ruminal Microorganisms, 83
Biotechnology of Animal Feeding, 88
Probiotics, 88
Biotechnology Crops as Animal Feed, 89
Biotechnology of Animal Products, 90
Some New Products made Possible by Biotechnology Include, 91

Chapter-3: Role of Biotechnology in Animal Health 93

Vaccines, 94
Veterinary v/s Human Vaccines, 95
Modern Veterinary Vaccines, 96
rDNA Technology to Prepare Vaccines, 98
Veterinary Viral Vaccines, 100
Live Vaccines, 100
Killed Vaccines, 101
Genetically Engineered Viral Vaccines, 101
Veterinary Bacterial Diseases Vaccines, 101
Veterinary Parasitic Vaccines, 102
Vaccines against cancer, 103
Swine Flu, 103
Foot and Mouth Disease, 103
Adenovirus Vector Vaccine, 104
Johne's Disease, 104
Diseases of Poultry, 105
Newcastle Disease in Poultry, 105
Avian Influenza (Bird Flu), 105
Rabies, 106
Disease Diagnostics, 106
The Nucleic Acid-hybridization Technique, 108
Diagnostics Based on PCR, 108
Diagnosis by Restriction Fragment Length Polymorphisms and Related DNA-based Approaches, 109
Diagnosis by DNA Probes and DNA Microarray Technology, 109
Antigen-Capture Enzyme-Linked Immunosorbent Assay (ELISA) Proteomics, 110
Disease Diagnosis Kits Developed, 110
Brucelosis, 110
Rabies, 110
Canine Distemper and Canine Parvo Virus, 110
Canine Hepatitis Virus, 111
Foot and Mouth Disease, 111
New Castle Disease, 111

Egg Drop Syndrome Virus-76 (EDS-76), 111
Proteomics for Disease Diagnostics, 112
Nanotechnologies in Diagnosis and Vaccine Development, 112
Monoclonal Antibodies, 113
Immunotherapy, 114
Problems, 114
Animal Genomics for Disease Resistance, 115
Gene Knockout Technology, 117
Gene Pharming, 117
Gene Therapy, 119
Types of Gene Therapy, 121
Germ Line Gene Therapy, 122
Somatic Gene Therapy, 122

Chapter-4: Modern Biotechnology Issues **123**

Environmental Biotechnology, 123
Global Warming and Animal Health and Production, 124
Genome Mapping, 126
Human Genome, 126
Livestock Genomics, 127
Cattle Genome, 127
Pig Genome, 129
Bioinformatics, 129
Major Branches of Bioinformatics Include, 130
List of Bioinformatics Software Commonly Used in Animal Genomics Research, 130
Sequence Analysis Software, 131
Genetics Analysis Software, 131
Functional Genomics Software, 131
Biofuels, 131
Advantages of Bioethanol, 133
Ecological, 135
Economical, 135
Looking to the Future: The Potential of Second-Generation Biofuels, 135
Future Role of Biofuels Depends on Profitability and New Technologies, 136

Chapter-5: Information Sources and Requirements for a Biotechnology Laboratory **139**

Information Sources for Biotechnology, 139
Books, 140
General Biotechnology Books, 140
Animal Biotechnology Books, 141

Review Articles, 143
Journals, 146
Animal Biotechnology Journals, 146
Biotechnology Job Listings, 147
The Scientist, 147
Patents, 147
Requirements for Setting up a Biotechnology Laboratory, 148
Essential, 150
Beneficial, 151
Useful, 152
PCR, 152
Specifications of a Good PCR Macine, 153
Gel Documentation System Electrophoresis, 153
Laboratory Safety and Hazards, 154

Chapter-6: Present Status and Future Prospects of Biotechnology in Animal Health and Production **157**

Checks and Balances in Biotechnology Applications, 159
Biosafety Issues in Animal Biotechnology Application, 159
World Organization for Animal Health (OIE), 160
Recombinant DNA Safety Guidelines (1990), 162
Foreign Trade (Development and Regulation) Act, 1992 (2006) (Draft Amendment), 162
Revised Guidelines for Research in Transgenic Plants and Guidelines for Toxicity and Allergenicity Evaluation of Transgenic Seeds, Plants and Plant Parts (1998), 162
Guidelines for Generating Preclinical and Clinical Data for rDNA Vaccines, Diagnostics and other Biologicals (1999), 162
The Food Safety and Standards Act (2006), 163
Transgenic Animals Bioreactors, 163
First Patent in Biotechnology, 165
Biotechnology - Ethical Concerns, 165
Some Priority Areas in the Field of Biotechnology, 166
Suggestions for Future Development of Biotechnology Activities in the Field of Animal Health and Production in India, 167
Epilogue, 167

Chapter-7: Biotechnology Research in India **169**

Animal Science Biotechnology Research Centres, 171
Current Status and Some of the Achievements in Animal Heatlh, 172
National Bioresource Development Board, 175
Animal Biotechnology, 176

Biotechnology Parks and Incubators, 176
National Institute of Immunology, New Delhi, 176
Bose Institute, Kolkata, 176
Jawahar Lal Nehru University, New Delhi, 176
National Centre For Cell Science, Pune, 176
Centre for DNA Fingerprinting and Diagnostics (CDFD) Hyderabad, 176
National Institute for Plant Genome Research (NIPGR)
New Delhi, 177
Institute of Bioresources and Sustainable Development, Imphal, 177
Institute of Life Sciences, Bhubaneshwar, 177
Bharat Immunologicals and Biologicals Corporation Limited, 177
International Centre for Genetic Engineering and Biotechnolgoy (ICGEB), New Delhi, 177
Animal Biotechnology Research Abroad, 178

Chapter-8: Techniques Used in a Biotechnology Laboratory **181**

r DNA Biotechnology, 181
Concept of r DNA, 181
DNA Preparation from Tissue, 183
Reagents, 184
Preparation, 184
DNA Buffer (Tris-EDTA), 184
RNase A (20 mg/ml), 185
Toe DNA Preparation for Southern Blots, 185
Southern Blot, 185
PCR, 186
Polymerase Chain Reaction, 186
The Cycling Reactions, 186
Denaturation at 94°C, 187
Annealing at 54°C, 187
Extension at 72°C, 187
DNA Sequencing, 187
The Sequencing Reaction, 187
Denaturation at 94°C, 187
Annealing at 50°C, 188
Extension at 60°C, 188
Separation of the Molecules, 188
Detection on An Automated Sequencer, 189
Assembling of The Sequenced Parts of a Gene, 189
Freezing of Buffalo Semen (Jindal, 1995), 189
Tissue Culture, 191

Preparation of Media, 192
Laboratory Procedure for IVF (Kharche *et al.*, 2008), 195
Source of Oocytes, 195
Preparation of Estrous Goat Serum, 195
Collection of Oocytes, 195
Laparoscopic Ovum Pick-up, 196
Oocyte Retrieval from Ovaries Collected form Slaughter House, 196
Slicing Technique, 196
Follicle Puncture Technique, 196
Follicle Aspiration Technique, 196
Follicle Isolation Technique, 196
Evaluation of Oocytes, 197
In vitro Maturation of Oocytes, 197
Preparation of Cells for Co-culture, 197
Assessment of Nuclear Maturation After *in vitro* Maturation of Oocytes, 198
Analysis of Nuclear Status is Ascribed to One of the Following, 198
Sperm Preparation for *in vitro* Fertilization of Oocytes, 199
In vitro Fertilization of Oocytes, 199
Assessment of *in vitro* Fertilization of Oocytes, 200
Assessment of Cleavage, 200
In vitro embryo culture, 200

References 203

Glossary 229

List of Noble Prize Winners 238

Index 243

1

Introduction to Biotechnology

World population is expected to increase from the current 6.3 to 7.5 billion by 2020. While spectacular rise in food production was made especially in Asia during 1970s and 1980s, the recent years have seen a slow down or even stagnation, raising concerns about food and livelihood security in the developing countries.

Biotechnology is perceived as a revolution throughout the world. Biotechnology has made the world a different place. Biotechnology is globally recognized as a powerful tool of plant and animal genetic modification (GM) that holds promise of improving productivity, profitability and sustainability of farm production systems, including those existing in small and poor farming situations. Animal production is on the threshold of this revolution waiting to reap immense benefits for the welfare of humankind.

Biotechnology is technology based on biology, especially when used in agriculture, food science, and medicine. Biotechnology in the modern prospective is the science of genetic manipulation of biological organism so as to engineer them for the production of better and more useful organisms. Biotechnology is a conglomerulation of several sciences encompassing the realms of biology as well as engineering. Humans have altered animals, plants, and even microbes through selective breeding for millennia, to the great benefit of society. Biotechnology increases our ability to continue this process at an accelerated pace and in a more directed manner. With these powerful tools comes an increased ability to do great good but there are risks and benefits to the technology. Although it is imperative that biotechnology be used responsibly and ethically, it is equally important that these powerful techniques for advancement not be sequestered due to unfounded fear of potential adverse consequences.

Biotechnology per se has been an integral part of human culture for a long time. Brewing, cheese making, baking etc. all rely on mastering fermentation using a variety of micro organisms and are as old as human history. Biotechnology although is a very old discipline but its modern connotation has came into offing since 1970 when the technology of gene isolation and injection have become a practical reality. Scientists are now working over time to prepare DNA libraries for the identification of useful genes and several such gene mapping. Atlases have been attempted for domestic animals.

The field of modern biotechnology is thought to have largely begun on June 16, 1980, when the United States Supreme Court ruled that a genetically-modified microorganism could be patented in the case of Diamond *vs*. Chakrabarty. Ananda Chakrabarty, working for General Electric, had developed a bacterium (derived from the *Pseudomonas* genus) capable of breaking down crude oil, which he proposed to use in treating oil spills.

The rise of genetic engineering, genomics, proteomics, and the creation of transgenic plant and animals has revolutionized activities as varied as brewing beer and the treatment of sewage and waste water to drug development, agriculture and animal husbandry. Biotechnology has achieved some dramatic advances in recent years in both crop and livestock production. These include transferring a specific gene from one species to another to create a transgenic organism; the production of genetically uniform plants and animals (clones); and the fusing of different types of cells to produce beneficial medical products such as monoclonal antibodies. Today, biotechnology has a number of applications in livestock production. It is being used to hasten animal growth, enhance reproductive performance, improve animal health and develop new animal products.

Feed, Growth and Production Biotechnology can increase the digestibility of low-quality roughage, and genetically modify plants to improve their feed value, such as the amino acid balance. The use of microorganism as a source of protein for human and animal food has been advocated to supplement normal protein sources. The growth of microorganisms is more rapid than that of the higher plants, makes them very attractive as high-protein crops; only one or two grain crops can be grown per year, while a crop of yeasts or moulds may be harvested weekly and bacteria may be harvested daily.The nutritional quality of silages is in part due to the high microbial content. Rapidly growing organisms such as bacteria and yeasts contain high uric acid than slow growing plants and animals. While the uric acid limits the daily intake of SCP for humans and monogastric animals such as pigs and chickens but ruminants such as cattle, sheep and goats can tolerate higher levels. The more important and difficult problem of waste fibre utilization requires microorganisms that can utilize lignocelluloses at very high rates. Although no such organisms are known for the present, biotechnology can help in the development of such microorganisms. Alkali treatment has been advocated as one method to increase the availability of ligno cellulose in straws to rumen microorganisms.

It can also provide enzymes, hormones and other substances that enhance animal size, productivity and growth rates. Synthetic hormone bST (bovine somatotropin) was among the first innovations available commercially. It can increase milk yield by as much as 10 to 15 per cent in lactating cows. Current development efforts are being made to investigate genes at a whole spectrum that affect growth and production within the animal. Ways to genetically engineer cattle to increase their own natural hormone production are being considered, thus eliminating the need for synthetic bST.

Biotechnology can greatly accelerate the the rate if a performance of a desirable characteristics (e.g. better growth rates, or increased milk production), which can be introduced into animal. Although classical breeding to enhance animal traits works well, it takes decades to produce major changes. While using biotechnology, an organism can be modified directly in a very short time if the appropriate gene has been identified.

A recent breakthrough in animal reproduction is the combined application of the existing in vitro fertilization, and the state-of-the-art of ultrasound-guided transvaginal oocyte pick-up (OPU) technique in cattle. When heifers reach puberty at 11-12 months of age, their oocytes may be retrieved weekly or even twice a week for embryo production and embryo transfer. There is even the possibility of applying this technology to juveniles. In this way, high-value female calves can be used for breeding long before they reach their normal breeding age.

Important areas which are likely to yield early dividends are the identification, characterization and cloning of the major histocompatible gene complexes those are expected to usher in an era of disease free domestic animals. The genetic engineering of rumen microbes is another area where production of high yielding domestic animals will be made possible. Production of monoclonal antibodies and other useful bio proteins from micro organisms is destined to revolutionize the science, of animal keeping in the world. Development of easy, fast-and reliable diagnostic kits for various diseases and for pregnancy diagnosis also holds promise. Animal health care is also set to be completely changed with the availability of biotechnology produced vaccines, for the treatment of various prevailing and emerging disease of animals. Vaccines against various viral diseases which are likely to be lot more effective and for diseases for which vaccines are non-exist today are likely to be available by the use of biotechnology. Although a significant beginning has been made in India but an accelerated pace is more than the need of the hour to reap the full benefits of this powerful technology.

The use of biomass as fuels help to reduce the greenhouse gas emission because the CO_2 released during combustion or conversion of biomass to chemicals, is removed from the environment by photosynthesis during the production of the biomass (i.e. plant growth). Biodiesel can be used as pure form or blended with petroleum based diesel fuel.

Worldwide, more than one-half of all biotechnology research and development expenditures are in the field of human health. At the experimental stage, a large number of drugs, diagnostic probes, vaccines etc. are frequently applied in livestock production prior to becoming available for use by humans. Developments in the pharmaceutical industry, therefore, have had considerable ramifications for animal production since many innovations in this area are also applicable to animals.

The biotechnologies affecting animal production can be defined under the three broad categories :

- Those biotechnologies which impose additional metabolic demand on the animals (e.g. biotechnologies affecting growth rate, milk yield or reproductive prolificacy).
- Those which alter the nature of the animal product but don't place additional metabolic demands (e.g. altering the composition of milk secreted or meat produced by domestic animals).
- Those which are designed to improve animal welfare (e.g. disease resistance).

The biotechnclogies can alternatively be grouped as follows.

- Genetically engineering of forage species, either to increase their productivity, or to improve their nutritional value.
- Genetically engineering of microorganisms to produce food/feed additives.
- Genetically engineering the gut micro flora.
- Production of therapeutic or prophylactic compounds-including vaccines. From genetically engineered microorganisms.
- Production of hormones or hormone analogues from genetically engineered microorganisms.
- Immunomodulation of physiological processes.
- Improved DNA based diagnostic procedures.

Biotechnology Products and Technologies

Biotechnology is one of the most progressively developing area worldwide, being it in research, development or entrepreneurship. The global turnover of biotechnological companies increased in 2007 to more than 55 billions USD, this is twice the 6.4% grow rate of worldwide pharmaceutical market (according to the IMS Health company that performed a market research). Most marketed biotech products are :

- Proteins available from gene farming.
- Recombinant vaccines
- Monoclonal antibodies
- Gene therapy
- Diagnostics

Biotechnology-derived proteins and polypeptides are a new class of potential drugs, which have commercial applications. For example, Insulin, used in the treatment of diabetes, was primarily extracted from slaughtered animals. Since 1982, human insulin Humulin(R) has been produced by microorganisms in huge fermentation tanks. The global sales of Humulin(R) and other biotechnology drugs like Epogen(R) (Epoten alpha from Aongen) and Reuombivlx HB(R) (Hepatitis B vaccine from Merck) have reached around 1 billion US $ of each. These products are convenient to make and more compatible with biological systems.

Bottlenecks in Fast Progress of Animal Biotechnology

Animal biotechnology has proceeded in two directions:

1. The production of animals for meat or milk
2. The creation of animals that produce biomedically useful proteins in their blood or milk.

The latter is a highly specialized application directed at human health and is limited to a few companies. Farmers and researchers have long used animal breeding to improve animals for food, but to date, genetic engineering has not been commercially used in animal breeding.

The production of high quality safe food, respecting animal welfare and the environment, are the primary objectives of animal husbandry. Particular focus is to be placed on the improvement of animal health and production. However, relevant genomics research is fragmented and the restructuring of the research capacity is required. Increased cooperation and knowledge sharing will bring together scientific excellence to make a real difference to animal and human health, and improve the quality of animal products.

Biotechnology research on farm animals is not widely conducted, and progress is relatively slow for several reasons.

- Relevant animal genomics research is fragmented and increased knowledge sharing and cooperation is required.
- Farm animals have much longer reproductive cycles than plants; so seeing the result of any breeding innovation takes much longer. Techniques for superovulation, ovum recovery, in vitro fertilization, nuclear transfer, cloning, and embryo transfer have

low rates of success; therefore, applying genetic engineering to animals is relatively inefficient.

- Animals are costly; so the low rates of success mean the economics militate against commercialization.
- Social concerns include "the inadvertent release of dangerous microorganisms, the safety of products derived from biotechnology, the impact of genetically engineered animals on the environment, animal welfare concerns, and our societal and institutional capacity to manage and regulate the technology and its products".

Why Biotechnology should be Used in Animal Husbandry and Dairy Production

Worldwide, almost 90% of the human food supply is provided by only 15 crop and eight livestock species. Introducing genes in the livestock species from microorganisms has been thought to be a means to ensure the continued productivity from these livestock species.

Animal Husbandry and Dairy Development plays a prominent role in the rural economy in terms of sustainable employment and the income of rural households, particularly, the landless, small and marginal farmers of India (Sharma and Mishra, 1987). It also provides subsidiary occupation in semi urban areas and more so for people living in hilly, tribal and drought-prone areas where crop output may not sustain the family. According to estimates of the Central Statistical Organization (CSO), the value of output from livestock and fisheries sectors together at current prices was about Rs. 2,25,841 crore at during 2005-06 (Rs. 1,85,166 crore for livestock sector and Rs. 40,675 crore for fisheries), which is about 31 per cent of value of output of Rs. 7,20,340 crore from total of Agriculture, Animal Husbandry and Fisheries sector. The contribution of these sectors to the total GDP during 2005-06 was 5.30%. India is endowed with the largest livestock population in the world. It accounts for 57 per cent of the world's buffalo population and 14 per cent of the cattle population. According to Livestock Census (2003), the country has about 18.5 crore cattle and 9.8 crore buffaloes.

Contribution of Livestock Sector to Food Basket

Agriculture including Animal Husbandry is the means of livelihood security for around two thirds of the work force of India. This makes it one of the most important sectors of the economy. At the time of independence, the revenue from the agricultural and animal husbandry sector was quite low compared to what it is today. The main reason for the increase in revenue is the increase in agricultural production that was brought out by the Green Revolution. Now we need another green revolution in the country to enhance the food and animal productivity possible by biotechnological revolution in the field of agriculature and animal husbandry.

The National Agriculture Policy aims to attain a growth rate in excess of 4 per cent per annum in the agriculture sector, stresses the importance of food and nutritional security issues and the importance of animal husbandry and fisheries sectors in generating wealth and employment. Since the present growth rate in crop production is around 2%, higher growth rates of 6 to 8% in animal husbandry sector would help in achieving the targeted growth rate of 4% for the Agriculture sector as a whole. The Policy proposes to accord high priority to increasing protein availability in the food basket and generation of exportable surpluses. Health care, fodder production, and freedom from animal diseases are some of the other areas of importance, as envisaged in the Policy document.

The contribution of livestock sector to the food basket in the form of milk, eggs and meat has been immense in fulfilling the animal protein requirement of ever growing human population. The present availability of animal protein in an Indian diet is less than 10 gm per person per day, as against a world average of 25 gm. However, keeping in view the growing population, the animal protein availability has to increase at least two fold, for maintaining the nutritional level of growing children and nursing mothers in India.

The country accounts for 14.2 per cent of the world's total milk production, retaining its position as the largest milk producer. Of the total milk produced in the country, 45-50% is consumed as fluid milk of which 18% is handled by organized sector both cooperative and industrial concerns. India possess 102 million milch animals and stand first in the world in total milk production.

The egg production in the country has reached 46.2 million numbers in 2005-06. The wool production in the country has reached 44.9 million kg during 2005-06.

Other Livestock Products: Livestock sector not only provides essential protein and nutritious human diet through milk, eggs, meat etc., but also plays an important role in utilization of non-edible agricultural by-products. Livestock also provides by products such as hides and skins, blood, bone, fat etc. which are raw material for several high value products.

India has enormous wealth of biodiversity in the form of domestic animals. Apart from being a source of economic upliftment of the livestock owners and farmers, these animals are profoundly associated with our social, cultural and traditional values. It is therefore, imperative that conservation, propagation and optimum utilization of the animal genetic resources available in the form of indigenous livestock wealth will not only provide food and financial security to the present generation but also enable us to meet the future challenges. Productivity of livestock needs to be increased to ensure a stable year-round supply of meat and dairy products.

The major livestock which will herald the biotechnology revolution in the future include cattle, buffalo, sheep, goat and poultry. Water Buffalo (*Bubalus bubalis* –(Linnaeus, 1758)) can rightly be called as India's Black Gold because of its importance as the key dairy animal of 21st century. It already has a major contribution to agrarian economy of India from livestock. Buffalo, apart from its use as dairy animal, it is also used in agricultural operations to pull carts with heavy loads eg sugarcane in India and other Asian countries. Their dung is used as manure for the soil and as a fuel when dried as cakes. Buffaloes in India contribute substantially to meat production as there is no religious taboo on meat consumption in non vegetarian population of all communities especially Hindus and Muslims, where consumption of beef and pork is a taboo, respectively. Asia is the home of the water buffalo, with 95% of the world population of which half is in India. Technological innovations are crucial for improving buffalo productivity during the 21st century. To identify opportunities and challenges specific to livestock production, the induction of innovative technologies are the biggest challenges which need to be studied and analyzed in detail before launching large scale programs.

World possesses 807 million goats and India with 124.4 million stands second after China. Goat population in the country is poised to reach 170 million by 2025. Goats in India contribute 14.7% population, which share 22% milk, 10.5% meat and 13% skins of the world and 23% population, 40 % milk, 14 % meat and 16.5 % skins of the Asian region. Around 89 % of the goats around the world are reared primarily for meat. About 47.5 million goats (40%) are annually slaughtered for meat in the country. Value of different goat produces in India works out to over Rs. 93500 million per annum. The goat sector generates about 4 % rural employment and 20 million families in India are engaged in goat keeping. Per capita meat consumption in India is less than 5 kg per year as against 13.7 kg in Pakistan, 38.6 kg in China and 58.6 kg in Brazil. Estimated annual demand for meat is 7.7 million MT against the availability of 5.6 million MT. Goats contribute 8.5% of the total meat and about 3.0% of milk production of the country. India exported 3 lakh live goats, 280 MT goat meat and 25 MT goat- skins and earned over 4 MT US $ during 2003. A great export potential exists for live goats, goat meat, goat skins and their products. Around 250 commercial goat farms have been established in different parts of the country. Thus the rapidly changing patterns of demand for livestock and livestock products, point to goat production being an important component of the agricultural economies of India.

The world's livestock sector is undergoing a massive transformation, a large part of which is heralded because of biotechnology. India and developing countries will have to meet increasing demands for meat and milk in the coming decades, with reducing methane emissions and environmental impacts. Animal biotechnology provides an opportunity to achieve these goals fruitfully and successfully in a shorter span to time. The major areas where such a strategy can be useful are :

1. Feed production and preservation technologies,
2. Feeding strategies,
3. Reproduction management strategies,
4. Disease control strategies,
5. Dairy and meat processing technologies.

The Government of India also envisages that the fields of biotechnological applications in the genetics and breeding, immunology and disease control should be some of the priority areas for livestock development in the country.

Biotechnology and Environment

Worldwide, agricultural activity, especially livestock production, accounts for about a fifth of total greenhouse gas emissions, thus contributing to climate change and its adverse health consequences, including the threat to food yields in many regions. Particular policy attention should be paid to the health risks posed by the rapid worldwide growth in meat consumption, both by exacerbating climate change and by direct contributing to certain diseases. To prevent increased greenhouse gas emissions from this production sector, both the average worldwide consumption level of animal products and the intensity of emissions from livestock production must be reduced. Environmental impacts of intensive animal production are partly associated with a mismanagement of animal excreta, leading to pollution of surface water, ground water and soils by nutrients, organic matter, and heavy metals. Climate change will cause significant changes in soil moisture, productivity and quality of pastures, liveweight of grazing cattle and species composition of native pasture.These issues have been discussed in detail in this book at appropriate places.

Biotechnology of Nutrient Partitioning

At various stages of an animal's life cycle, certain tissues achive metabolic priority over other tissues of the body. Examples are metabolic requirements of the fetus and placenta during pregnancy and mammary gland during lactation. To accommodate the increased requirements of high-priority tissues, the animal must alter the pattern with which nutrients are partitioned among the tissues. This shift in nutrient partitioning is called as **homeorhesis** and is associated with altered physiological states requiring a shift in the division of the circulating nutrient pool. The most dramatic example of homeorhesis is the changes that occur with the onset of lactation. The metabolic requirements of lactation far exceed maintenance requirements of the adult animal. Many metabolic adaptation have evolved in various species to assist the dam in meeting the metabolic changes. Metabolic adaptations are most necessarily under endocrine control. Studies on hormone concentration in serum with milk yield has shown the role of certain hormones in crossbred cattle and buffaloes and the association was studied through the use of correlation coefficients (Jindal, 1989, Jindal and Ludri, 1990 a,b, c, 1993, 1994, 1995). The role of growth hormone, insulin and thyroid hormone was elucidated during lactation. Singh and Ludri (1998, 1999, 2000) and Yash Pal and Ludri (1997) elucidated the role of prolactin during early lactation in goats. Growth hormone is galactopoietic which is evident from numerous studies where growth hormone administered to animals increased milk yield upto 25%. Bauman and his colleagues have tried to explain the process as a homeorhesis phenomenon i.e. the nutrients are partitioned more towards milk production in case of growth hormone injected animals. Use of external hormones like bST can increase the partition of nutrients towards production purposes but the availability of hormones through natural means is a constraint. The biotechnological availability can alleviate this situation. The availability of bovine growth hormone or somatotrophin through biotechnology opened a new avenue to control the homeorhesis in lactating animals towards more milk production. The chronology of bovine somatotrophin studies are given as below.

- 1936 Russian scientists reported that injecting dairy cows with crude bovine pituitary extracts of bST increased milk yield (Asimov, *et al.*). However, wide-spread commercial use of the extracts was never seriously pursued since only very small and impure amounts were obtainable from cows at slaughter houses.
- 1950s scientists injected pituitary extracts of bST in U.S. children with the hope of treating hypopituitary dwarfism, supplemental bST did not stimulate growth and had no effect on humans.
- 1970s Recombinant DNA technology was developed, leading to commercial bulk production of bST.
- 1979 Professor Dale Bauman at Cornell University conducted the first study in which high-producing cows were supplemented with pituitary bST.
- 1982 Recombinantly produced human insulin was introduced. Which was identical to natural human insulin and was made by a process similar to that used for bST.
- 1982 Professor Bauman at Cornell University conducts and reports results from the first study in supplementing cows with recombinant bST.

In 1993 Monsanto Company received U.S. FDA approval for POSILAC bovine somatotropin. FDA Commissioner David A. Kessler, M.D., said "There is virtually no difference in milk from treated and untreated cows. In fact, it's not possible using current scientific techniques to tell them apart. We have looked carefully at every single question raised, and we are confident this product is safe for consumers, for cows and for the environment. This has been one of the most extensively studied animal drug products to be reviewed by the agency. The public can be confident that milk and meat from bST-treated cows is safe to consume."

In 1999 Animal scientists published a study supporting the effectiveness of POSILAC in on-farm settings over a four year period. The Northeast Dairy Herd Improvement (DHI) study involved 27,000 cows in 340 northeastern commercial dairy herds with cows supplemented with POSILAC and herds with cows that were not. The scientists concluded supplementation with POSILAC significantly and consistently increased milk production throughout multiple years of use. The average days in milk and average age of the herd remain consistent between control herds and the herds supplemented with POSILAC. In India, Ludri and his colleagues conducted studies on lactating buffaloes and found that use of bST increased the milk yield significantly depending on the dose. However, the availability of bST (recombinantly derived somatotrophin) is common in many countries of the world, but India, it is yet to be approved for commercial use.

Genetic Programming for Better Animals

Likewise, selection of better animals i.e., high producing animals which are genetically programmed to partition more nutrients towards production purposes like milk production. In future, the genetic programming is likely to be altered to achieve goals of greater productivity from our livestock to feed the ever bulging human population of the world. Genomics is powerful tool for unravelling the molecular basis of phenotypic variation in domestic animals. This also offered additional information for genetic improvement of efficiency and quality traits with the identification of single loci of major effect on variation between animals. Genetic improvement could be accelerated for yield and other economically important traits by directly selecting upon the genetic differences underlying the phenotypes. During the domestication, the goat has undergone intense natural selection processess for various phenotypes. Selection in this species has led to distinct phenotypes associated with meat, milk, fibre production thriving in tropical environments in some part and tolerating specific pathogens. These selective pressures have differentiated sub-populations and produced phenotypes according to the need of the region. Therefore, it is necessary to utilize molecular markers to select high performance individuals for suitable environment for enhancing productivity and sustainability in goat production. It is necessary to combine molecular markers and production traits in an efficient manner for attaining higher productivity. DNA marker information, which identifies important allelic variation within the genome, could be incorporated into genetic evaluations to provide producers with selection tools that increase the rate of genetic improvement for lowly heritable traits.

Biotechnology of Reproduction

We also need to produce more milk and milk and also make our livestock to reproduce as per our wishes, so that the goals of more productivity can be achieved. There are several biotechnologies which can control and augment reproduction and they are discussed in detail in the next chapter. New reproductive bio-techniques will go a long way for improvement and conservation of livestock breeds, increasing the prolificacy of the less prolific livestock breeds, reduced kidding interval and age at sexual maturity and has opened new vistas for application of biotechnological tools for improvement of domestic animal industry in the country in general. Some of the important reproductive biotechniques are artificial insemination, embryo transfer, in vitro maturation, sexing, cloning, parthenogenesis, transgenesis etc. Some of these techniques are discussed at appropriate places in the book in detail. Cryo-preservation of gametes has provided an extremely useful tool for improvement of livestock at a much faster rate. Cryopreserved semen and embryo offer a unique tool for improving the genetic potential of livestock including cattle, buffalo, sheep, goat, pig, horses, camel and other species useful to human being. The present day breeds of livestock are a result of continued and sustained selection of superior animals over a period of several thousand of years. For faster genetic improvement cryo-preserved gametes provide excellent opportunities.

Biotechnology of Animal Health

Biotechnology of animal health is primarily focused on two major thrust areas i.e., diagnostics and vaccinology besides a host of related issues like gene therapy and gene pharming. Besides these two major issues there is a possibility of producing diseases resistant farm animals through genetic manipulation of the major Histocompatibility Gene complex. In case of diagnostics, the PCR offers potential advantages over other conventional tests. The risk of false negative is limited in PCR test. Further, cell culture tests are less sensitive due to the presence of inhibitors like interferons and presence of enzymatic inhibitors. The conventional PCR is being replaced by Real Time PCR due to several advantages like better precision, high resolution and automatic assays. The disadvantage of real time PCR is that it is very expensive and is not suitable for field use. The possibilities are immense. Many of these issues are discussed in the chapter on biotechnology in animal health improvement.

2

Biotechnology in Animal Production

Genetic potential of animals has been exploited by man ever since domestication of livestock took place. Human ingenuity has brought about considerable improvement in animal productivity from time to time. This has been further helped by man's urge to explore newer frontiers of science and to evolve newer technologies for application to have greater gains. The recent developments in biotechnology have opened up exciting possibilities for a rapid increase in productivity of domestic animals through its applications. Animal biotechnologies can help increase in animal productivity in several ways namely

- By increasing the production of animal products by promoting growth and increasing nutrient intake efficiency.
- By increasing the reproduction rate of domestic animals.
- By enhancing the quality of animal products (for example lean meat)

Biotechnology is being applied to enhance animal production. The major areas under which this can be brought out are:

- Reproductive biotechnologies.
- Use of bST (growth hormone)
- Livestock genomics and marker assisted selection (R.F.L.P.)
- Livestock transgenics

Reproductive Biotechnology

A number of biotechnological methods are being developed to increase the reproductive potential of livestock. These include:

Artificial Insemination (AI)

Especially since the development of efficient semen freezing methods, AI has become the most widespread biotechnology applied to livestock production. By allowing for the widespread use of small numbers of elite sires (Proven bulls). AI has had a dramatic impact on selection intensity. AI has allowed for the implementation of progeny testing scheme in cattle or buffalo production which has had a major impact on the improvement of the herd by increasing the accuracy of selection. Semen production in the country has increased from 22 million straws (1999-2000) to 44 million straws (2007-2008) and the number of inseminations has increased from 20 million to 40 million.

History and Future Perspectives of Artificial Insemination

Artificial insemination (AI) was the-first great biotechnology applied to improve reproduction and genetics of farm animals. It has had an enormous-impact worldwide in many species, particularly in dairy cattle. Pioneering efforts to establish AI as a practical procedure were begun in Russia in 1899 by Ivanow. Earlier method of collecting-semen was from sponges placed in the vagina of mount animals. Later Milovanov designed and made practical artificial vaginas and other items, many similar to those used today. A Danish "invention" of importance in popularization of AI was the straw for packaging semen. Later Cassou (1964)-produced straws which were used commercially and became famous as French straw. Especially since the development of efficient semen freezing methods, AI has become the most wide spread biotechnology applied to livestock and especially cattle production. It has been estimated that approximately a fifth of the breedable female population in the world is now bred by artificial insemination (Thibier and Wagner, 2002). AI remains as one of the most important assisted reproductive technologies. The three cornerstones for its application are: it is **simple, economical and successful**. The most widely used test of sperm quality from the-initial stages of AI development until the present time-has been the assessment of the proportion of normal,-progressively moving sperm (Foote, 2002). In addition to examining sperm with brightfield microscopes, differential interference contrast microscopes, multiple stains, flow cytometry, and computer assisted-sperm analysis (CASA) have contributed to improved quantification of sperm motion. Rapid optical density methods for measuring sperm concentration have replaced tedious hemocytometric procedures. Fertility of sperm is the ultimate test of sperm quality. Many-tests of semen quality in addition to motility and morphology, such as the hypoosmotic swelling test, mucous or gel penetration, and integrity of the DNA have been correlated with fertility. The first major improvement in the-AI procedure was the development of a yolk-phosphate semen extender (Phillips and Lardy, 1940). Salisbury *et al.* (1941) improved the media by buffering the egg yolk with sodium citrate. The next major stimulus to AI of dairy cattle was an improvement of about 15% in fertility resulting from a better method of initially protecting sperm from cold-shock (Foote and Bratton, 1949) and the control of some venereal diseases by the addition of antibiotics (Almquist *et al.*, 1949; Foote and Bratton, 1950). AI with the use of antibiotics added semen eliminated venereal diseases, reduced embryonic death, and achieved high fertility. Later Tris-buffered egg yolk-glycerol also provided excellent protection for sperm either frozen or unfrozen (Davis *et al.*, 1963). Initially semen doses were stored in solid carbon dioxide at -79°C. Researchers later demonstrated that sperm survival at -196°C was virtually infinite, whereas biological changes occurred with storage at -79°C. Frozen semen is less fertile than fresh semen (Shannon and Vishwanath, 1995), as many semen additives to improve fertility of frozen semen have been tested with minimal success. However, a recent report (Amann *et al.*, 1999) provided preliminary support for a peptide that increased fertility when added to frozen thawed semen.

Freeze drying of sperm has not been reported to be successful. Wakayama and Yanagimachi (1998) demostrated the preservation of the intigrity of DNA of freeze-dried sperm in Microinjection Technique. However, Because of the difficulty of insemination, general management and low value per animal, AI, particularly of goat and sheep, is not widespread (Foote, 2002). AI in buffaloes has been adopted quite successfully in India thanks to several researches carried out at CIRB, NDRI, NDDB and other centres (Reviewed by Jindal and Chopra, 1993, Jindal, 1998). Advantages which can be achieved in a successful artificial insemination programme can be seen from table below, where the national average (kg) of milk production per cow in USA more than doubled during a span of 25 years by judicious use of artificial insemination technology combined with better feeding and management (Foote, 1981).

Table : Milk production per cow during the past 25 years in USA (National Average (Kg))

1950	2415
1955	2655
1960	3195
1965	3772
1970	4431
1975	4706
1979	5227

(Foote, 1981)

Semen Cryopreservation

Freezing of spermatozoa needs a concept of sperm preservation from body temperature to freezing and-thawing, because these are phases which cryopreserved sperm must successfully negotiate to retain fertility. If we measure efficiency of freezing sperm of different domestic animals by degree of fertility considering only the bovine has freeze preservation reached fertility levels comparable to that of liquid semen inseminations or natural service. In all-other species fertility results are not optimal or at least-not comparable to natural service or artificial insemination (AI) with fresh semen. Species differ ences are-seen considering the different steps of freezing. There exists a large difference concerning cooling sensitivity in the range between body temperature and refrigeration-temperature, boar sperm being the most sensitive to chiling. Most methods described are only effective when used in conjunction with gamete and embryo freezing methods. In addition cryopreservation plays a crucial role in conservation programmes aimed at maintaining genetic diversity.

The history of cryopreservation of gametes; No meaningful introduction to history of cryopreservation can be without the mention of Anton van Leeuwenhoek who discovered sperm in 1677 using magnifying lens. Italian physiologist L. Spallanzani in 1780 discovered the fertilizing power of sperm resided in the sperm carried by the spermatic fluid. When the semen was filtered, the liquid that passed through was sterile, but the residue on the filter was high in fertilizing power. He further observed that freezing stallion semen in snow or winter cold did not necessarily kill the "spermatic vermiculi" but held them in a motionless state until exposed to heat, after which they continued to move for seven and a half hours (1803). Philips and Lardy (1940) brought forth egg yolk as a semen dilutor to protect the sperm cells against damage during cooling. Almquist *et al.*, (1949) introduced the addition of the antibiotics penicillin and streptomycin to semen to control pathogenic microorganisms.

A good extending medium is key to the successful process of cryopreservation of semen as well as embryos. The following characteristics are essentially required for a good extender and it should

1. *It should* provide nutrients as a source of energy.
2. *It should* protect the spermatozoa against the harmful effects of rapid cooling.
3. *It should* provide good buffering capacity to prevent harmful shifts in pH as lactic acid is formed.
4. *It should* maintain the proper Osmotic pressure and Electrolyte balance.
5. *It should* be able to inhibit bacterial growth
6. *It should* increase the volume of the semen so that it can be used for multiple inseminations.
7. *The diluents should* be easy to prepare, low in cost and give a clear picture of spermatozoa or embryos under microscope and render no problem in the cleaning of glassware and other containers.

8. *It should* also ensure a long shelf life of spermatozoa /embryos on storage.

Principles of cryopreservation: It is well known that low temperature storage ensures longer keeping quality of any material including foodstuffs. Refrigeration storage of food material is a common knowledge. The storage at 4^{0}C or lower temperatures like deep freeze delays the harmful bacterial growth, which spoils the food or other material. This applies to preservation of gametes as well. Therefore, earlier attempts to extend diluted semen at refrigeration temperature extended the life span of spermatozoa by few days. This period was sufficient for transportion of semen up to a few hundred kilometers for insemination of cattle at a distant place. However, the quality soon deteriorated. Freezing at below refrigeration temperature resulted in death of spermatozoa or embryos because of formation of ice crystals. As you might be aware that ice occupies more space than the water from which it is formed resulting in bursting of the vessel in which it is contained. Therefore, the availability of cryoprotectants, which prevent the formation of ice crystals during freezing and developments of such procedures which cause minimum damage to the cells during freezing became key to such long term preservation of gametes.

The availability of cryoprotectants is very useful to prevent the damage from cold shock. Glycerol is the most commonly used cryoprotectant although ethylene glycol and dimethy sulfoxide are also considered good.

Egg yolk is known to protect bull spermatozoa against cold shock. However, egg yolk is not a suitable constituent for extenders used for preserving goat semen, because the presence of an egg yolk coagulating enzyme, reported to be present in goat semen.

Artificial insemination technique based on the principle of cryopreservation originated before the Second World War in several European countries. By 1945, the use of artificial insemination was not a successful a tool for the genetic improvement but it did get cows in calf. Therefore, for progeny testing of dairy bulls artificial insemination became an important scientific tool. However, it was soon realized that it took so long to accomplish 'proof' that most of the bulls tested had long since disappeared from the scene. A major development which took place at that time revolutionized the dairy industry around the world in coming years. Polge *et al.*, (1949) discovered a practical method for the long term preservation of the semen of certain species by deep freezing to temperatures of -79°C by means of dry ice. In fact, this was a demonstration that semen can be preserved in frozen state. Later developments in the field of semen freezing lead to the worldwide acceptance and use of this technique in artificial breeding programmes resulting in considerable improvement in milk yield of livestock species.

Buffalo

Buffalo semen freezing is essential to achieve our aim of improved productivity from domestic/indigenous buffaloes. Semen freezing can enhance the use of superior quality high genetic merit buffalo bulls on more number of buffalo cows and thus accelerate the pace of genetic improvement considerably. Bhattacharyya and Srivastava (1955) and Roy *et al.*, (1956) carried out initial experiments on deep freezing of buffalo semen in India.

The cost involved in setting up a semen freezing laboratory as well as problems encountered with freezing and lower conception rate obtained sometimes, are major deterrents to the fast spread of semen freezing and its use.

Goat

Frozen semen use of goats has not been as successfully as in cattle. Most of the procedures developed for cryopreservation and insemination of cattle semen has been extrapolated for use with goat semen. However, the efficacy of several of them for goat spermatozoa has not been proved beyond doubt. Therefore, the use of AI in goats is limited because of a variety of management problems, prejudice, economics and less well developed technology.

The importance of AI is to utilize outstanding sires over larger population. Application of this technology controls the spread of venereal diseases and extensive use will reduce the cost of maintenance and housing of breeding sires either on small holdings or in organized farms. It is fairly established that artificial insemination is a successful technique in goat production, but its fruitful exploitation on a large scale has not been possible due to various reasons.

1. Shortage of males of superior genetic makeup of indigenous breeds.
2. Small size of flock having 1 or 2 goats in a family. They are not able to keep breeding sires exclusively for getting their goats pregnant.

3. Artificial insemination in goats requires technical personnel and equipment.
4. The conception rate by AI is not satisfactory in many cases.

Artificial Vagina (AV) is the most commonly accepted method for semen collection of bulls, buffalo bulls, bucks, horses etc. Artificial Vagina suited and designed for different animal species are commercially available.

It is very difficult to collect semen completely free from microorganisms even under strict hygienic conditions but washing of preputial sheath with 0.9% sterile saline solution, 20 minutes prior to semen collection significantly reduces the bacterial load in semen. Presence of egg-coagulating enzyme phospholipase secreted from bulbo-urethral gland of buck in semen prevents successful storage of semen at 5°C in yolk containing diluent (Roy, 1957). Buck semen could be stored successfully up to 48 h at 4 to 7°C with satisfactory sperm motility and fertility. However, it is difficult to maintain satisfactory motility and fertility beyond 48 h of storage. Recentresion in indicated that goat semen extended to the diluents having antioxidants (3 mM vitamin C, 1.5 mM vitamin E, 5 mM glutathione oxidized and 5 mM glutathione reduced respectively) can be stored up to 96 hours at refrigeration temperature for further use as liquid semen for AI as well as transportation to any place for immediate use (Jindal *et al.*, 2009). Ravi Ranjan *et al.*, (2009) found that Marwari buck semen extended in the diluents having 10% egg yolk level could be stored upto 72 h at refrigeration temperature for further use as liquid semen for AI.

Buck semen is highly susceptible to cold shock. When semen is frozen and kept at very low temperature (at -196°C), the metabolic reactions of spermatozoa are completely arrested which makes possible to preserve semen for a prolonged period. Seminal plasma has been blamed as a limiting factor of goat semen freezability and fertility (Roy, 1957). During the process of freezing, spermatozoa gets damaged either due to the formation of internal ice crystals or to increased solute concentration in the media, or the interaction of both of the above physical factors (Salisbury *et al.*, 1978). Efforts for deep freezing of buck semen by conventional dry ice alcohol method (-79°C) started as early as in 1950 (Smith and Polge), but the real success in cryo-preservation of buck semen was achieved 1970s by use of washed semen with liquid nitrogen vapour technology. Experiments have been carried out in different parts of the world to standardize the various steps involved during deep freezing of buck semen with different dilutors, cooling rate, equilibration period, glycerolization, post thaw techniques etc. Purdy (2006) reviewed the recent attempts on cryopreservation of goat semen.

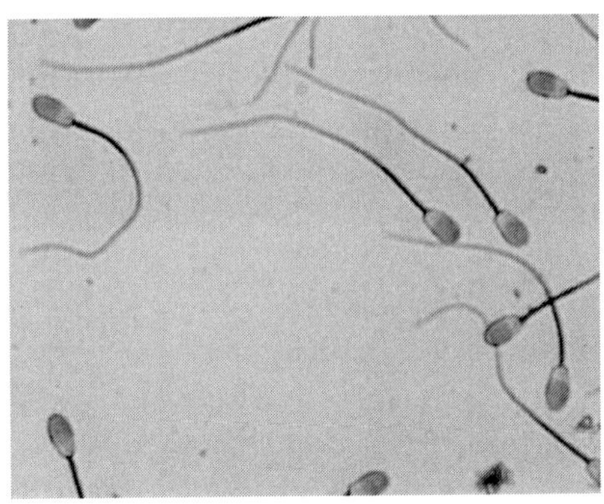

Goat sperm acrosome for quality evaluation

The deep freezing at -196°C of buck semen with different dilutors has been studied in relation to freezability and fertility by many workers. The results of recent goat sperm cryopreservation research are highly variable (Purdy, 2006). Frozen-thawed goat sperm motility, viability, acrosomal integrity and fertility differ dramatically between reports. Therefore, research should focus on obtaining a better understanding of goat sperm physiology so that assays can be developed to better assess the sperm cell quality and its fertility following cryopreservation. Various levels of glycerol (0 to 9%) have been tried as cryoprotective agents for freezing buck semen but 6% glycerol seems to be best for freezing of goat semen. Equilibration is essential for spermatozoa extender interaction which minimizes sperm damage during freezing. There has been loss of sperm motility in yolk containing diluent due to presence of yolk coagulating enzyme in buck semen (Roy, 1957). Straw freezing is now accepted as the most suitable method for freezing buck semen. Sperm motility after freezing varied with thawing temperature and time.

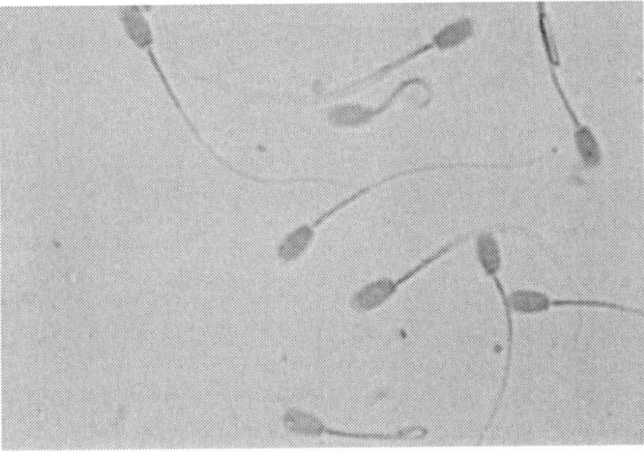

HOS Test for goat sperm showing coiled and non coiled sperms

The definition of "successful" cryopreservation for mammalian sperm is a relative term, particularly when different species are compared. Dairy bulls have been selected for the ability of their sperm to "freeze well" and consequently dairy bull sperm repeatedly have high percentages of motile, viable sperm cells after freezing that are capable of fertilizing oocytes. On the other hand, sperm from bucks, rams, boars and stallions survive freezing less consistently and therefore are potentially less fertile. Significant affect of additives and diluent osmolarity on post-thaw sperm motility has been observed.

Cryoprotection and Anti Oxidants

Cryopreservation is associated with the production of reactive oxygen species which lead to lipid-peroxidation of sperm membranes. It has been demonstrated in human, bull, and mouse spermatozoa that cryopreservation is associated with oxidative stress. Moreover, freezing and thawing of bovine spermatozoa increase the generation of reactive oxygen species (ROS), causing DNA damage, cytoskeleton alterations, inhibition of the sperm–oocyte fusion and affecting the sperm axoneme that is associated with the loss of motility. Natural antioxidants exert a protective effect on the plasma membrane in cryopreserved bovine spermatozoa, preserving both metabolic activity and cellular viability. Enzymatic antioxidant defense mechanisms in seminal plasma and spermatozoa include superoxide dismutase, catalase and both glutathione reductase and peroxidase activities. Among non enzymatic antioxidants there are urate, ascorbic acid, vitamin E, taurine, hypotaurine, carotenoids, pyruvate and glutathione (GSH). The cryopreservation procedure deplete both seminal plasma and cellular antioxidant systems; in fact, seminal plasma is either removed or highly diluted during freezing and cellular antioxidants are lost during freezing and thawing process.

Cryopreservation and Genetic Stability of Livestock Spermatozoa

Development of techniques for the successful freezing of spermatozoa has progressively evolved over the past many years. Cryopreserved semen has been utilised in the artificial insemination of livestock species, even though the detrimental effects of cryopreservation on sperm function and fertility are known to some extent. A significant decrease in sperm motility has been observed in bulls (Foote and Parks, 1993), dogs (Olar *et al.,* 1989), goats (Salamon and Ritar, 1982) and humans (Zavos *et al.,* 1995) after sperm cryopreservation. Pregnancy rates from AI are also affected after cryopreservation of stallion (Samper *et al.,* 1991), bull (Valcticel *et al.,* 1994) and goat spermatozoa (Ritar and Salamon, 1983) as compared to fresh semen insemination. In addition to sperm motility, various aspects of sperm morphology have been shown to be affected by cryodamage during freezing. For example, the concentration of acrosome reacted spermatozoa is significantly increased after freezing and thawing (Valcticel *et al.,* 1994). Cryopreservation is also responsible for membrane damage of bovine (Valcticel *et al.,* 1994) and human (Mahadaven and Trounson, 1984) spermatozoa. A decrease in the percentage of morphologically normal spermatozoa in the ejaculate has been correlated with lowered fertility in caprine, (Chandler *et al.,* 1988), bovine (Saake, 1972) and equine species (Jasko *et al.,* 1990). Hence, the decrease in normal sperm morphology resulting from cryopreservation may be responsible for the corresponding decrease in fertility of these samples. While freezing of spermatozoa has been found to affect aspects of sperm head morphology, the classification of spermatozoa was based on visual evaluation methods. In the assessment of human sperm morphology, these methods have been shown to be difficult to apply between technicians and the results are highly variable (Davis *et al.,* 1995).

The sperms used for artificial insemination often shows signs of deterioration; in other cases, the semen characteristics seem positive, yet pregnancy proves impossible to obtain. Ensuring the integrity of spermatozoa, DNA is necessary to obtain a pregnancy through artificial insemination. Male factors have been shown to have an effect on embryo development. (1), As illustrated by the guidelines of the World Health Organization for standard semen analysis (2), count, motility, and morphology have historically been used as indicators of male fertility potential. However, the values between proven fertile and infertile have a large overlap; thus, routine semen analysis is not a reliable indicator of fertility potential or pregnancy outcome. Elevated levels of sperm DNA fragmentation are now a well-recognized factor for negative pregnancy outcome (Erenpreiss *et al.,* 2006), and there is increasing evidence that sperm nucleus integrity also should be routinely examined. Sperm genome anomaly is often one of the factors involved in failure to obtain an embryo and/or pregnancy (Filatov *et al.,* 1999). Sperm DNA fragmentation may be caused by internal factors, such as apoptosis or the production of free radicals by the spermatozoa (Sakkas *et al.,* 1999, Shen *et al.,* 2002), or external factors such as leucocytes (Zorn *et al.,* 2000). One absolute requirement for the use of cryopreserved sperm is that the cryopreservation process must not induce alterations that would impair progeny development. Altered spermatozoa are known to produce reactive oxygen species (ROS) very rapidly (Alvarez and Storey, 1984; Aitken and Fisher, 1994), in particular after cryopreservation (Alvarez and Storey, 1992).These peroxidation products are highly deleteritous for DNA and can induce both strand breaks and base modification (Lopes *et al.*, 1998). The consequences can be dramatic since it has been previously demonstrated that sperm exposure to DNA damaging agents altered the genome of progeny, as well as progeny viability and development. The impact of the DNA fragmentation on the fertilization rate (Lopes *et al.*, 1998) and pregnancy rate have been highlighted (Benchab *et al.*, 2003,Bungum *et al.*, 2004). Three major methods are generally used to evaluate sperm DNA integrity: sperm chromatin structure assay (SCSA) (Evenson *et al.*, 1999), the terminal deoxynucleotidyl transferase-mediated. The sperm used for assisted reproduction techniques (ART) often shows signs of deterioration; in other cases, the semen characteristics seem positive, yet pregnancy proves impossible to obtain. Ensuring the integrity of spermatozoa DNA is necessary to obtain a pregnancy in good condition through ART.

Digoxigenin-dUTP nick-end labeling (TUNEL) technique, and the comet assay method. For the SCSA technique, 30% appears to be the consensus pathologic threshold. For the TUNEL technique, the pathological threshold is currently unclear, oscillating between 12% and 36.5% (17, 18). Sergerie *et al.*, place the threshold value for fertility at 20%, a figure that also had been proposed by Benchaib *et al.* The Sergerie study, which compared fertile men with infertile men, showed that fertile men have a lower percentage of spermatozoa with fragmented DNA than infertile men; however, it did not answer the question what threshold value is relevant for a population of infertile men entering an ART protocol.The range of the pathological threshold of the percentage sperm DNA fragmentation seems to be between 15% and 20%. The majority of studies carried out on sperm DNA fragmentation and ART outcome have examined semen that was used for ART procedures. The stability of sperm DNA fragmentation values in comparison with standard semen parameters allows for a potential role in predicting ART outcome.

Reliable assays to assess the stability of goat, buffalo or cattle semen are highly desirable in cryobiology, which involves molecular techniques, are efficient for evolutionary, ecological and population genetics studies. Among these a technique termed as Random Amplified polymorphic DNA- polymerase chain reaction (RAPD-PCR) has been successfully applied in genetic studies of various animals species

(Chapaco *et al.*, 1992) as well as for molecular characterization of bovine populations. RAPD-PCR is based on amplification of DNA in the polymerase chain reaction (PCR) by short random primers. This technique has several advantages over others. Oxidative stress has negative influence on sperm DNA and proteins (Davis, 1987). DNA fragmentation appears to be inversely correlated with semen quality, particularly sperm count, morphology and motility. Moreover negative correlations have been observed between the stability of DNA in the sperm nucleus and the fertilizing capacity of spermatozoa *in vivo* and *in vitro* (Aitken *et al.*, 1998).

Insemination Technique

Cattle and Buffalo

Insemination technique in cattle and buffalo is well standardized and poses not much of a problem.The rectovaginal technique of artificial insemination in cattle and buffaloes is one of the most commonly used methods. First the excess fecal matter is removed from the rectum. Subsequently vulva is wiped clean and dried with an absorbent paper towel to prevent contamination. The cervix is located and insemination gun containg the semen inserted per vagina.The cervical os is probed with the gun tip and semen deposited by pressing the plunger.The technique is very simple in the hands of a trained inseminator. Efficient and accurate detection of estrus are essential in dairy herds using artificial insemination. Inaccuracies in estrus detection result in insemination of cows that are not in estrus, thus lowering the herd conception rate.

Sheep and Goat

Insemination in small ruminants like sheep and goats, is difficult and needs special skills and training. Insemination in sheep and goat should be done by lifting the hind legs of females towards sun at an angle of 60° from the ground. Insemination crates in which the hind quarters of the doe or ewe are elevated should be preferred. A speculum with a light sources and a catheter is required. Locate the slit like opening of the cervix with the help of glass speculum and inseminating gun fitted with sheath. Gun and sheath should be introduced inside the cervical fold as far as possible. It has been observed that in parous if the female is in proper heat, sheath is passed through the cervical fold to body of the uterus giving better fertility results. Remove the speculum and gun with sheath together. Separate sheath is used for individual animal. In absence of the sunlight, head light should be used. The technique can be employed in the farmer's flock at their door using frozen semen of superior/proven sires but a trained inseminator is required.

Conception Rate with Artificial Insemination

Reproductive efficiency depends on the fertility (conception rate) and fecundity. Conception rate is the percentage of cows that have become pregnant through artificial insemination (AI) or natural mating. If you inseminate ten cows and you get six calves, then your conception rate will be 60%. In general, conception rate is influenced by the breed, age, and nutrition and management practice. In addition, conception rate depend on the timing and number of inseminations. Nutrition and semen-handling have a major effect on conception rates. Low conception rates can also occur when faulty insemination techniques are used. Inseminating cows or buffaloes not in heat, inseminating the wrong cow or inseminating too early or too late, will seriously affect conception rates. Rough handling of semen straws can reduce conception rates. Treatment of cows with HCG or GnRH normally does not improve conception rate in normal cows except in special cases having problems of repeat breeding. Inseminating too early reduces the conception rate by reducing the sperm viability and number of sperms at the site of fertilization.

Relatively low conception rate has been reported with chilled and frozen buck semen. It is believed that frozen-thawed sperm are less motile and lack stamina to transverse the highly viscous cervical mucus, but phagocytosis of the sperm by leukocytes is also considered as a cause of the reduced fertility. Conception rate with frozen buck semen varied between 35 to 65 percent. Difference in fertility may be due to age, breed, and stage of heat and insemination technique. Conception rate in goat was poor due to deposition of frozen semen over the opening of cervix. Improvement in fertility in buck semen insemination was observed due to deep cervical insemination technique.

In sheep, surgical (laparoscopic) AI gives much better pregnancy rates than cervical AI. However, laparoscopic AI is more laborious and also more invasive than cervical AI. Molinia *et al.* (1996) showed that the difference in pregnancy rates between surgical and non-surgical AI was even larger with frozen semen compared to fresh semen: 20% versus 70% pregnancy with 180×10^6 vs. 10×10^6 sperm.

Multiple Ovulation and Embryo Transfer (MOET)

Embryo transfer biotechnology generated lot of interest among people around the world during the past thirty or more years. This biotechnique enables achieving a greater number of offspring from selected females than was possible by applying traditional means in animal reproduction. By increasing the number of offspring that can be obtained from monotocus species in particular, MOET has the potential to genetic improvement by enhancing the selection intensity on the female side. In cattle, however the species in which this technology is again the most widely disseminated. The major impact of MOET might result from the reduction in generation interval vis-a-vis the conventional progeny testing scheme, if sires are selected based on the performances of their MOET produced full sister rather than the performance of their female progeny: the so called MOET nucleus scheme. Embryo transfer created the basis for the emergence of such fields as mincromanipulation of embryos and genetic engineering. Embryo transfer also made possible examination of mechanism regulating early development of pregnancy.

Embryo transfer techniques have been perfected for majority of livestock species like cattle, buffalo, sheep, goat, pig and horses.

Despite associated technical hurdles, MOET has the potential to play an important role in developing countries, where the implementation of a large scale AI based progeny testing scheme would be difficult to implement.

- Second generation of reproduction biotechnologies
- Super ovulation of superior females (donors)
- Artificially inseminating them
- Transfer of fertilized eggs into surrogate mothers (recipients)
- Female fertility is multiplied

Advantages of Embryo Transfer

- Rapid multiplication of the elite female breeding stock
- Sire evaluation through progeny testing
- Conservation and preservation of breeds
- Creation of disease free herd
- Economical transport of livestock
- Salvage of reproductive function
- Future applications through research

Embryo transfer technique consists of three steps

- Superovulation
- Collection of embryo either surgically or non surgically
- Transfer of embryos to suitable recipients

Most recent developments which made embryo transfer a successful proposition include

- Non surgical recovery and transfer
- Cryopreservation of embryos

Applications of Embryo Transfer

- Amplify reproductive rates of valuable females
- Especially useful in species with low reproductive rates and long generation intervals.
- It is possible to obtain offspring from genetically valuable cows that have become infertile due to injury, disease, or age by means of superovulation and embryo transfer.

MOET Includes the Following Steps

Identify genetically valuable animals accurately by progeny testing so that the best can be used as parents of the next generation

Since half of the genes come from the male, it is extremely important to use genetically superior bulls.

Cows are more preferable as recipients as compared to heifers because of less difficulty at calving. Moreover it is more difficult to transfer embryos in heifers as compared to cows.

Superovulatory Treatments

Superovulation can be defined as increased ovulatory response by external hormone therapy, above a level that would be expected to occur normally. Mammalian ovary contains many more oocytes than are destined to ovulate. Several factors affect the process of superovulation like rate of growth of follicles, entry of dominant follicles into the pool of growing follicles and rate of loss of follicles through atresia.The two generally accepted methods of superovulating cattle are based on two different gonadotrophins, although there are many minor variations of these methods. The simplest is to give an intramuscular (i.m.) injection of 1800–3000 IU (usually 2000–2500 IU) of pregnant mare's serum gonadotrophin (PMSG), more correctly designated equine chorionic gonadotrophin (eCG), followed by a luteolytic dose of prostaglandin F2 α or an analogue i.m. two to three days later. A second dose of prostaglandin is often injected 12 to 24 hours after the first, and seems to improve embryo production.

The second method of superovulation is to give eight to ten injections of follicle stimulating hormone (FSH) subcutaneously (s.c.) or i.m. at half-day intervals.

There are a number of factors which influence the superovulatory response which include age, breed, repeated superovulatory treatments and hormone variability. The source of the FSH also determines the superovulatory response. For example in buffaloes, FSH-p from porcine source is most effective whereas the FSH-o from ovine source or FSH-e from equine source were much less effective. This may be because of the difference in sequence homology as well as FSH:LH ratio in the source hormone. The dosage, route and frequency of administration is also important factor in determining the superovulatory response.

Some of the superovulatory hormones used in livestock are

- Super-OV
- Folltropin
- FSH-P
- Folligon- PMSG
- Trophovet-PMSG

Insemination

If liquid semen is available, between 10 to 50 × 10^6 motile spermatozoa are inseminated 12 hours after the donor is observed in oestrus and a similar quantity 12 hours later. If frozen semen is used, one ampoule or straw each time is inseminated 12 and 24 hours after the donor was first noticed in oestrus.

Recovery of Embryos

Prior to 1976, most bovine embryos were collected via mid-line laparotomy or, less commonly, via a flank incision. In that year, several groups published efficacious methods for non-surgical (transcervical) recovery of embryos. In most cases, embryos are recovered six to eight days after the beginning of oestrus (day 0). Recovery procedures are carried out by manipulation per rectum.

Non-surgical recovery is carried out with the help of Foley catheter.

Fluid from the uterus is usually collected in either 2-litre graduated cylinders or through a 75 μ mesh filter (Embryo collector). The flush fluid is examined for embryos under a microscope. Embryos are then evaluated for their quality. Embryos collected six days post-oestrus should be post-compaction or so-called tight morulae.

In cattle, embryos are routinely transferred to the uterine horn. The most commonly used instrument for non-surgical transfer is the standard Cassou inseminating gun for French straws.

Synchronization of Recipients

There are many methods of synchronizing reproductive cycles of recipients to match those of donors. In some circumstances, natural synchrony is feasible, but in most cases some recipients will need to be synchronized to augment those whose oestrous cycles match the donor's naturally. The most widely accepted procedure for synchronizing recipients is administration of a luteolytic dose of prostaglandin F2 α or a suitable analogue during the luteal phase. This is probably superior to using natural oestrous cycles. Injecting potential recipients with two doses of prostaglandin at 11 day intervals when stages of the reproductive cycle are unknown also works well if cattle are cycling.

Cattle

First claim of successful transfer of bovine embryo was made by Willet *et al.*, (1951) by surgical method. The first successful non-surgical transfer in bovine was reported by Mutter *et al.*, (1964). Since then numerous workers have attempted and studied embryo transfer in bovine and other species.

Application of embryo transfer to the cattle industry began in the early 1970s when European dual-purpose breeds of cattle became popular in North America, Australia and New Zealand. Breeders and speculators sought means to circumvent the high costs and lengthy quarantine periods linked to the importation of European breeding stock and to capitalize on premium prices that progeny from these rare dams and sires could command. More than half a million (538,312) bovine embryos were reported to have been transferred in 2002, more than half of them (52%) after on-farm freezing and thawing and 15% of them having been produced *in vitro* (Thibier, 2003). North America is still the centre of most activity (35% of the transfers) but this is static or declining in contrast to South America, where ET is expanding and accounted for 22% of the world's transfers in 2002. Europe and Asia each reported about 17% of the total number of transfers in 2002. There is general agreement that a severe limitation to the more widespread use of ET is the problem of reliably inducing superovulation in selected donors.

Buffaloes

The research on embryo transfer in buffaloes got big boost with the report of a successful non surgical transfer of bubaline embryos by Drost *et al* (1983)using techniques of embryo transfer in cows. Chantaraprateep (1989) reported the birth of first successful embryo transfer calf in swamp buffaloes. Cruz *et al.,* (1991) reported the successful transfer of a Murrah buffalo embryo into Phillipine swamp buffalo. The buffalo ovary has been shown to have smaller population of recruitable follicles at any time than the ovary of bovine. Several laboratories latter reported limited success in getting successful embryo transfer in buffaloes (Mishra, 1993, Jain *et al.,* 1988, Madan, 1992, reviewed by Jindal *et al.,* 1994). There are differences between cattle and buffaloe which cause the results to be less successful in buffaloes. Superovulation is less predictable in buffaloes than in cows. Embryo development is more rapid in the buffalo. The consequence of this fact is that buffalo embryos must be collected between five and six days after the onset of oestrus (versus day 7 in the cow). Corpus luteum of the buffalo is already smaller and more deeply embedded (therefore more difficult to palpate) than that of the cow.

Goats

Detrimental effects of repeated surgical embryo-recoveries cause adhesions and reduce fertility of donor females, limiting the usefulness of the embryo transfer technique for genetic improvement. Two-effective methods of embryo recoveries that require less surgical trauma are laparoscopic and other-nonsurgical procedures (Amoah and Gelaye, 1997). Ovulation profile in Jamunapari goats indicated that ovulation occurred in this breed during 32-36 hours following estrus. (Goel *et al.,* CIRG). Multiple ovulation and embryo transfer (MOET) in goats has not become widely used due to its unpredictability (Baldassarre and Karatzas, 2004).

Jamunapari goats started cycling from May and their oestrous activity reached peak in September through December. Oestrous activity started declining from January and complete cessation in March-April. Oestrous duration ranged 24-48 hours. Incidence of short cycles and long cycles also has been observed in goats. The goats and sheep embryo transfer has to be surgical because of the small reproductive organs which cannot be manipulated per rectum as in case of cattle and buffaloes. Using surgical methods for embryo transfer at CIRG, the superovulatory respone in Jamunapari goats averaged 6.80 ovulations/goats, embryo/ova recovery 4.50/goat and transferable embryos 4.0/goat (Goel *et al.*, 2003-2004). Goel *et al.* (2003-2004) attempted the non surgical method of embryo transfer in goats but with limited success rate.

Horses

The first foals after embryo transfer were born practically at the same time in Japan and England during the late spring in 1973 (Tischner, 1987).The first successful trials at cross species embryo transfer were carried out between mare and she-asses, and she-asses with mares (Allen, 1982).

In vitro Maturation, fertilization and culture of Oocytes

Recent progress made in the field of oocyte maturation and *in vitro* fertilization of farm animals has provided a new dimension for exploitation of animal reproductive set up. This technology has great potential for application on genetically superior animals to have progeny from it in numbers many fold than numbers it would produce in its life time, if left to reproduce normally. Breifly this technique involves collection of oocytes from the ovary, *in vitro* maturation of oocytes, *in vitro* capacitation, fertilization and culture of the embryos so produced.

Historical Development

- Rabbit — Chang, 1959
- Golden hamster — Yanagimachi and Chang, 1963
- Mouse — Whittingham, 1968
- Rat — Toyoda and Chang, 1974

- Dog — Mahi and Yanagimachi, 1976
- Human — Steptoe and Edwards, 1978
- Cattle — Brackett *et al.*, 1982
- Pig — Chang *et al.*, 1968
- Lamb — Chang *et al.*, 1968
- Tiger — Donoghue *et al.*, 1990
- Horse — Zhang *et al.*, 1990
- Buffalo — Madan *et al.*, 1991
- Goat — Keskintape *et al.*, 1994

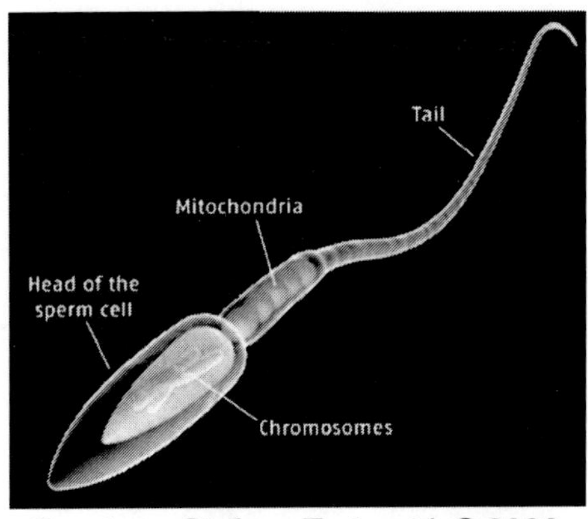

Courtsey: Stefano Tartarotti, © 2006

Chang (1959) was the first to achieve successful *in vitro* fertilization in rabbits. When the same technique was applied successfully nearly two decades later to human gametes (Steptoe and Edwards, 1978), the term *in vitro* fertilization or IVF became widely popular. Brackett *et al* (1982) produced the first IVF bovine calf which encouraged the animal scientists to apply this unique technique in other species of farm animals. Mammalian ovary contains hundreds of thousands of oocytes; the number of progeny a female can produce in its life time is small. In farm animals, the number of times a female can become pregnant is severely limited by the extended duration of gestation. The large pool of unused genetic material contained in the mammalian ovary can be utilized fruitfully by resorting to IVF-ET. IVF can prove to be an invaluable technique not only for the supply of more number of embryos from valuable animals, the production of viable embryos through IVM/IVF is much cheaper than its production by MOET (Woolliams and Wilmut, 1999). IVF can be helpful in progeny testing and testing of several sires simultaneously. It can also be made use of in case of genetically superior animals having irrepairable impairment of the reproductive tract. For most of the IVF studies slaughterhouse material is used.

In vitro production of embryos by IVM/IVF require basic manipulations such as oocyte collection, and IVM, oocyte activation and *in vitro* culture of zygotes to a certain stage before they can be transferred into the uterus of synchronous recipient. Appropriate developmental and trophic signals orchestrated to ensure normal development.

The number of embryos that can be obtained from a cow/year using MOET is on an average limited to the order of 20 or less, the development of oocyte harvesting in conjunction with IVM and IVF increases this number by a factor of at least 5. Moreover OPU, can be applied to pregnant animal as well as prepubic animals. The impact of these methodologies on genetic response operates through the same channels as MOET i.e. increase of selection intensity on the female side and increase of selection accuracy on the male and female side.

In Vitro Fertilization (IVF), an assisted reproductive technology (ART) in which one or more eggs are fertilized outside a female's body. This technique has been used extensively in animal embryological research for decades, but since 1978 it has been successfully applied to human reproduction. In human reproduction, the process involves stimulation of the growth of multiple eggs by the daily injection of hormone medications. (It is also possible to conduct IVF without the use of the hormone medications; a single egg would develop and be retrieved). The eggs are recovered by one of two methods: sonographic egg recovery, the more common of the two, which uses ultrasound guidance to retrieve the eggs, or oocytes; or laparoscopic egg recovery, in which retrieval is made through a small incision in the abdomen. Gamete intrafallopian transfer (GIFT), the technique is similar to IVF, but the harvested eggs and sperm are placed directly into the Fallopian tubes, with fertilization occurring in the woman's body. In zygote intrafallopian transfer (ZIFT), the procedure is similar to GIFT, but the beginning-stage embryos (zygotes) are placed directly in the Fallopian tubes.

Recovery of immature oocytes is a prerequisite for the *in vitro* maturation. Early attempts in farm animals or human beings were surgical or laparoscopic (Morgensterm and Soupart, 1972, Lenz and Lauristen, 1982) either by giving local or general anesthesia, with or without treatment of donors with suitable gonadotrophins. Results of a series of experiments suggest that because of the physical characteristics of the pre-ovulatory cumulus mass, it was considered possible that needle with somewhat larger inside diameter would be more useful. Now reports are available even on oocyte retrieval on commercial basis. The oocytes recovered by ultrasound guided needle, the number of oocytes aspirated per cow averaged 6.3; FSH treatment 3 days before aspiration increased the oocyte recovery rate to 8.3 (Looney *et al.*, 1994). Where slaughter house ovaries become the source of oocytes, two methods, dissection and aspiration of gollicles for collection have generally been used. Depending upon the species, 18 to 21 gauge needles have been used by different workers (Younis *et al.*, 1991, Deb and Gaswami, 1990). The number of oocytes obtained from cow ovary averaged 16.0 out of which 79.8 % were suitable for culture.

In Vitro Maturation of Oocytes

Ovum maturation in mammals is accompanied by resumption of meiosis. This is initiated during prenatal life or shortly after birth. The oocyte reaches the diplotne stage of prophase just before or immediately after birth. At this stage, by a mechanism yet not fully understood, the meiotic process is arrested with the nucleus (germinal vesicle) in prophase I of meiosis (Guraya, 1977). Meiotic maturation refers specifically to the process of nuclear progression from the diplotene or dictyate stage of first meiotic prophase to metaphase ll. Upon appropriate stimulation from circulating pituitary gonadotropins (FSH and LH) or in some species, after removal from graffian follicles and culturing *in vitro*, dictyate stage oocytes resumes meiosis leading to final maturation (Linder *et al.*, 1980).

In almost all studies of mammalian *in vitro* oocuyte maturation, the basic medium has been supplemented with serum or crystalline albumin. Addition of serum during maturation of primary oocyte is reported to prevent zona hardening and enhance the potential of oocyte and development after fertilization (Epig and Schrroeder, 1986). Findings of Lu *et al.*, (1987 a, b) suggest that serum may contain factors that promote the acquisition of developmental competence of oocytes druing maturation *in vitro*. There exists a controversy regarding the use of FCS and serum from an animal in estrus or at different stages of estrous cycle. Fukui and Ono (1989) used FCS and estrus cow serum. They reported that FCS was better than estrus cow serum for oocyte maturation. Minate and Toyoda (1982) reported that resumption of meiosis was inhibited by pig serum which contains germinal vesicle breakdown inhibiting factor. Murzamadiev *et al.* (1983) in their study on sheep oocyte maturation found that highest percentage of oocyte maturation (63%) was achieved with serum taken on day 16 of estrus cycle as compared to day 3 and 10. The use of serum in culture media is done with an idea to provide an environment as near as possible to *in vivo* conditions.

In cattle, use of 20 % estrus serum in maturation medium TCM - 199 has been shown to be capable of yielding 90 % oocytes at metaphase -ll . Gordon and Lu (1990) used FCS to avoid fluctuations in hormonal levels in adult animals and to give uniform treatment. Serum from different stages of estrus cycle has been tried by different workers to provide natural level of hormones required for maturation of oocytes.

Follicular fluid in which the oocytes remain embedded before aspiration is also sucked with oocytes. It is quite controversial whether it promotes or inhibits maturation of oocytes, if added to maturation media. Follicular fluid was reported to be ineffective in preventing resumption of meiosis in sow (Leibfried and First, 1980). Purine nucleotides present in follicular fluid help in oocyte maturation and attainment of competence. The replacement of fetal calf serum or buffalo estrus serum with follicular fluid gives comparable results during oocyte matruation (Yadav *et al.,* 1997).

Temperature is one of the important factors which influence the resumption of meiosis in both intra and extra follicular cultures. Katska and Smorey (1985) studied the influence of culture temperature on in vitro maturation of bovine oocytes. Bovine oocyte removal from the follicles 2 to 6 mm diameter were matured *in vitro* for 20 h at 33, 35, 37, 38° and 39°C. The percentage of oocytes in metaphase-ii increased from 2.8 at 33°C to 56.1% at 35°C and 73 at 37 to 39°C. Eng *et al.* (1986) reported that pig oocyttes cultured at 39°C had a higher percentage of polar body formation than did those cultured at 37°C. From these studies one can conclude that any working termperature in the range of 37 to 39°C would be most suited for *in vitro* maturation of oocytes.

In vitro Sperm Capacitation

When scientists first attempted *in vitro* fertilization, the sperms would not fertilize the oocytes. Sperms recovered from male reproductive tract would not fertilize but when recovered from female tract, they would. Spermatozoa of mammals required to be present in the female reproductive tract before they become capable of fertilizing (about 12 hours in cattle). Many studies (Austin, 1951; Chang, 1959) confirm that the environment of female reproductive tract changes the spermatozoon conferring on it the ability to penetrate and fuse with the egg. Capacitation can be achieved *in vitro* by addition of certain chemicals to the *in vitro* system. As *in vitro* environments are based largely on empiricism rather than on precise knowledge of embryo needs, they invariably provide a sub-optimal environment. Spermatozoa of mammals required to be present in the female reproductive tract before they become capable of fertilizing. According to Bedford (1983) capacitation could be defined as the concurrent changes that respectively allow the acrosome to react to physiological levels of trace calcium ions and cause tail to express beat pattern that results in hyperactivated motility. In vitro capacitiation of spermatozoa has been achieved by incubation of epididymal and ejaculated spermatozoa in media ranging from complex tissue culture media to simple balanced salt solutions supplemented with energy source and blood sera, serum albumin and female reproductive tract fluids (Bondioli and Wright, 1983; Anand *et al.,* 1989). Anand *et al.* (1989) used Krebs ringer bicarbonate medium containing pyruvate and lactate as energy source, 3 h incubation for epididymal and 4 h for ejaculated spermatozoa of bucks was reported necessary for capacitation and acrosomal reaction to take place. Calcium was shown to be an essential requirement which was needed for motility maintenance. Iritani *et al.* (1984) also used Krebs ringer bicarbonate for *in vitro* capacitation of bull spermatozoa. They suggested that ejaculated bull spermatozoa can be capacitated in a chemically defined isotonic medium. Role of albumin has been indicated for *in vitro* capacitation of mouse spermatozoa (Aonume *et al.,* 1982; Go and Woolf, 1985) and for cattle sperms too the same thing has been suggested by Byrd (1981). Andrew and Bavister (1988) induced acrosome reaction in hamster spermatozoa by using a chemical capacitation medium. However, such sperms could not enter the hamster eggs. Zona pellucida of hamster eggs required for fusion with sperms the presence of albumin in the chemical medium during sperm egg co-incubation. The above report indicates that albumin appears to be important for capacitation of spermatozoa *in vitro*. Albumin acts as a lipid solubilizing protein and dissolves the lipid layer of acrosome. It acts as a sterol acceptor during acrosome reaction (Go and Woolf, 1985). Heparin, calcium and caffeine are most commonly used for *in vitro* capacitation of spermatozoa. These are used in combination as well as separately, irrespective of species. The percentage of spermatozoa increased exponently with time during a 4 h incubation when spermatozoa were incubated with heparin. Chikamatsu *et al.* (1989) treated bovine spermatozoa by giving swim up treatment in media containing heparin or caffeine and calcium ionophore. Fertilization rate was higher with caffeine and calcium ionophore treated spermatozoa. Noland and Olson (1989) observed that $CaCl_2$ inducted temporal sequence of events, viz desruption of sperm rouleaux, loss of refractibility by tthe apical segment of the spermatozoa. Ijaj and Hunter (1989) demonstrated that treatment of bovine spermatozoa resulted in capacitation by washing in bovine serum albumin, saline and incubation for 4 h in calcium ion free Tyrode's medium. Dodds and Seidel (1984) studied the effect of caffeine, calcium ion and capacitation time on IVF in mice. Caffeine concentration of 0.9, 2 and 6 mM resulted in mean fertilization rate of 50.1, 58.8 and 65.4 % respectively. Spermatozoa capacitated in medium with 1.8 mM calcium ions fertilized more ova than when no calcium ions were present. The most efficient acrosome reaction took place when the medium contained both a monovalent ion Na+ and Ca++, simultaneously. El-Manufy *et al.* (1986) studied the effect of different levels of caffeine to unwashed ejaculated spermaozoa, which significantly increased sperm motility. In the presence of 2.8 mM caffeine, spermatozoa maintained their initial motility for at least 2 h at 37°C. The maximum stimulation of fructolytic activity was obtained with 2 mM, whereas minimum stimulation was found with 8 to 10 mM caffeine. Epididymal spermatozoa appeared to be more influenced by caffeine than ejaculated spermatozoa. Data on epididymal sperm motility demonstrated that caffeine (2 to 10 mM) significantly stimullated and maintained initial motility for at least 3 h at 37°C. Marks and Ax (1985) observed a significant difference in the binding affinity for heparin between bulls of high and low fertility. Calcium, caffeine and heparin act through metabolic pathways and then heperactivate the motility of sperm cell. This hyperactivity utilizes the reserve food in sperm cell and decrease the motility after a particular time. The hyperactivation regulates the sequence of activities for acrosome reaction. IVF has been used as a method to rank the fertilizing ability of bulls. This technique in case of bucks is yet to be tested.

- First IVF calf "Parthama" has revealed successful utililization of this technique in buffalo in India (Madan *et al.*, 1991).
- Scientists of Central Institute for Research on Goats Makhdoom, Farah, Mathura (UP) successfully produced an in vitro fertilized (IVF) goat kid in August 2006 (Kharche *et al.*, 2008).

Embryo transfer technology has become a tool for enhancing lifetime productivity of livestock. A large number of embryos are required for the success of Embryo Transfer programme under field condition. With variable superovulation response, reduced embryo recovery rates and the interruption of the normal reproductive cycle of the donor, *in vitro* maturation and *in vitro* fertilization procedure is an attractive alternative for the production of large number of embryos required for the rapid multiplication of superior germ plasm of goats with lowered fertility (eg. Jamunapari goat), upgrading a commercial or purebred breeding programme and conservation of endangered breeds of goat. Laparoscopic ovum pick-up is of major importance in the production and early propagation of such goats.

Secondly, a large number of early and late stages of embryos are required for the biotechnological experiments such as micromanipulation and microinjection for the production of identical twin, sexed offspring, cloning and transgenic goat. However the number of embryo obtained by superovulation is limited and expensive due to unpredictable superovulatory response; reduce embryo recovery rate and limitations of repeated surgical embryo collection in small ruminants. Therefore the use of ovaries collected from slaughter house material as a source of oocytes for IVM, IVF and embryo development is an attractive alternative for large scale economical production of early and late stages of embryos that can be used in the development of new biotechnologies.

Scope

- Multiplication of superior germ plasm in Cattle and goat through IVP and ET
- To evaluate fertilizing ability of spermatozoa
- Efficient use of semen from valuable bull
- To increase offspring from selected females
- To reduce generation interval in livestock

- To provide synchronously developing pronuclear stage ova for nuclear transfer or gene injection
- To preserve the germ plasm in the form of embryos
- To overcome reproductive failure in human through the use of intracytoplasmic sperm injection (ICSI) and gamete intra fallopian tube transfer (GIFT).

Limitations

It has been suggested that good buffalo oocytes with more than three to five cumulus layers recovered from large-sized follicles in cold seasons when cultured in TCM-199 supplemented with serum, follicle-stimulating hormone, and cysteamine resulted in maximum maturation rate and subsequent embryonic development after insemination (Suresh *et al.*, 2009). The above study is for buffalo oocytes but is expected to be true for other species too.

- There are increase frequencies of abortion, dystocia and perinatal death.
- The success rate is limited to 5 to 10% of calves born out of total oocytes collected in cattle whereas in human 25 to 28% pregnancies out of total embryos transfer.
- Congenital malformation increases to a level of 3.7% as against 0.8% in normal pregnancy in cattle.
- Chromosomal abnormalities increase from 27 to 45% IVP embryos in cattle.
- In human mean gestational length as well as mean birth weight increases as compare to normal pregnancy.
- In human 45% of IVF babies have neonatal problems as compare to 7.6% in normal cases.

Embryo Cryopreservation

Embryo cryopreservation is a means of long term storage of valuable strains, breeds or animals. Embryos are preserved in a cryoprotectant then slowly frozen. Embryos are then transferred to liquid nitrogen tanks for long-term storage.Cryo-preservation of gametes has provided a extremely useful tool for improvement of livestock at a much faster rate. Cryopreserved semen and embryos offer an unique tool for improving the genetic potential of livestock including cattle, buffalo, sheep, goat, pig, horses, camel and other species useful to human being. The present day breeds of livestock are a result of continued and sustained selection of superior animals over a period of several thousand of years. For faster genetic improvement cryopreserved gametes have excellent opportunities.

After the first successful cryopreservation of sperm cells in 1949, attempts were made to cryopreserve other cells and tissues. Whittingham (1971) reported first successful freezing and thawing of mouse embryos. Wilmut and Rowson (1973) reported the first calf from a deep frozen embryo. Table below shows the first reports of successful freezing of mammalian embryos.

Mouse	Whittingham, 1971
Cattle	Wilmut and Rowson, 1973
Rabbit	Bank and Maurer, 1974
Rat	Utsumi and Uuhara, 1974
Sheep	Willadson *et al.*, 1976
Goat	Bilton and Moore, 1976
Horse	Slade *et al.*, 1985
Cat	Dresser *et al.*, 1988
Buffalo	Techakumphu *et al.*, 1989

The procedure for cryopreservation includes initial exposure to and equilibration with the cryoprotectants, cooling to sub zero temperatures, storage, thawing and finally dilution and removal of the cryoprotectant with return to a physiologic environment that will allow further development. Cells must maintain structural integrity throughout the cryopreservation procedure. Storage is usually at -196°C, the temperature of liquid nitrogen. Above a temperature of -80°C, cells gradually lose viability, but below -130°C, there is insufficient energy for most reactions. Step wise dilution procedure is sometimes preferred as compared to single step method for removal of cryo-protectant.

All oocytes and embryos suffer considerable morphological and functional damage during cryopreservation. The extent of the injury depends on factors including the size and shape of the cells, the permeability of the membranes, and the quality and sensitivity of the oocytes and embryos. The purpose of cryopreservation procedures is to minimize the damage and help cells to regenerate. Almost all cryopreservation strategies are based on two main factors: cryoprotectants and cooling warming rates.

Cryopreservation of oocytes and embryos is a crucial step for the conservation of animal genetic resources. However, oocytes and early embryos are very sensitive to chilling and cryopreservation and although new advances have been achieved in the past few years, the perfect protocol has not yet been established (Pereira and Marques, 2008). All oocytes and embryos suffer considerable morphological and functional damage during cryopreservation but the extent of the injury as well as differences in survival and developmental rates may be highly variable depending on the species, developmental stage and origin (for example, *in vitro* produced or *in vivo* derived, micromanipulated or not). Currently, there are two methods for gamete and embryos cryopreservation: slow freezing and vitrification. We have experienced both techniques but vitrification has become a viable and promising alternative to traditional approaches especially when dealing with *in vitro* produced or micromanipulated embryos and oocytes. Recently new strategies based on emerging studies in the field of lipid research have been used to reduce intracellular lipid content in bovine *in vitro* produced embryos and therefore increase their tolerance to micromanipulation and cryopreservation. The addition of a conjugated isomer of linoleic acid, the trans-10, cis-12 octadecadienoic acid to embryo culture medium more than twice improved embryo post-thawing viability after micromanipulation and vitrification.

Details regarding the factors affecting cryoprotection of embryos are discussed in the review (Kumar *et al.,* 1998), that application of cryobiological principles can result in high survival rates for morulae and blastocysts in several livestock species; however, an optimal method for one species is not always applicable to embryos of any other species. Recently, technologies have been developed to cryopreserve pig embryos, notorious for their extreme sensitivity to cooling; horse embryo cryopreservation is in its infancy. In the near future, use of preserved embryos could be a routine breeding alternative for all livestock producers providing

1. Preservation methods for maternal germplasm,
2. Global genetic transport,
3. Increased selection pressure within herds,
4. Breeding line regeneration or proliferation, and
5. Methodology for genetic rescue.

Ultrasonography in Reproductive Management

Pregnancy diagnosis is an important tool to measure the success of a reproductive management, to allow for early detection of problems and to achieve resynchronization of nonpregnant cows. Pregnancy diagnosis is usually done by rectal palpation. The main disadvantage of rectal palpation is that it cannot be performed until later in gestation than some other methods. Usually rectal examinations take place between 45 and 60 days after insemination in cattle. The use of ultrasonography has further increased the reliability of the pregnancy diagnosis. At about day 26 of pregnancy in heifers and day 28 in cows, pregnancy can be diagnosed accurately under field conditions by ultrasonography. However, Anwar *et al.* (2008) reported that using real time B-mode ultrasound scanner equipped with a 3.5 MHz probe, 100% accuracy of pregnancy diagnosis was achieved on day 42 of gestation in sheep.

Ultrasonography is a diagnostic technique that involves directing high frequency sound waves at tissues in the body to generate images of anatomical structures. Ultrasonography is also called sonography, diagnostic sonography, and echocardiography when it is used to image the heart. Ultrasonography has become immensely popular and useful in reproductive biomedicine. Use of embryo transfer technology has successfully resulted in establishment of pregnancy in many domestic animals; however, the diagnosis of pregnancy is still a problem. Ultrasonography has been a great help for accurate diagnosis of pregnancy in recipient animals. Ultrasound technology allows *in vivo* estimation of carcass composition. The use of ultrasonography in embryo transfer program is useful for the :

- Evaluation of the super-ovulation response in donor cows
- Diagnosis of pregnancy in recipient animals.
- Visualization of fetal sex.

Advantages of ultrasonographic imaging techniques

- The high reliability of the results that are generated by ultrasonography
- Pregnancy diagnosis may be conducted relatively early after insemination (i.e., as early as 25 days after insemination).
- Ultrasonography techniques are non invasive and are therefore painless.
- They are applicable to live animals.

- Another advantage of ultrasonographic imaging over the existing X-ray technology is that it is useful for examination of soft tissues of the bodies where X-ray failed.

The basic principle of ultrasonography is that it uses high frequency sound waves produced by transducer to produce images of soft tissues and internal organs. The internal organs and tissues have inherent abilities either to transmit or reflect the sound waves to varying degrees.The portion of the sound waves that is reflected is received by a transducer, converted to electrical impulses and displayed on the ultrasound screen as a series of dots. The characteristics of the particular tissue determine what portion of the sound beam will be reflected. Liquids donot reflect sound waves, therefore the image of a liquid containing structure appears black on the screen (eg follicle filled with follicular fluid will appear as a black dot). On the other hand, dense tissue like bone, cervix etc reflect much of the sound beam and appear white on the screen. Other tissues appear in various shades of grey depending upon their echogenicity or ability to reflect the sound waves.

Ultrasonographic examination of cattle for pregnancy diagnosis is as given below.

The procedure for ultrasound examination in cattle and buffaloes essentially include restraining the animals and evacuation of the rectum because faecal material can cause distortion of the ultrasound image.After removal of the fecal material, the ultrasound transducer is covered with conducting gel and inserted in the rectum. The cervix is located and the uterine body is examined in longitudinal planes as the transducer is further inserted. When the junction of the uterine horn is reached, the tranducer is moved along the dorsal surface of the uterine horn so that the horn is examined in oblique planes, the transducer is then rotated laterally to examine the ovary. The palpation or manipulation should not be attempted while doing ultrasonic examination. Animals are examined while restrained in a crush. The contents of the rectum are carefully emptied using a lubricated, gloved hand prior to insertion of the probe. The ultrasound probe is then carefully inserted into the rectum. The reproductive tract is located and the probe is moved over the uterine horns and body. The operator may determine pregnancy status, stage of gestation and number of foetuses. The operator may examine the uterine fluid, the embryo/ foetus-for viability and size and any other relevant features of the tract or conceptus. The ovaries may be located and examined-for presence and size of normal follicles, corpora lutea or other significant structures (Pierson and Ginther, 1988).

Ultrasonography in small ruminants ie goat and sheep is very popular as a method of pregnancy diagnosis. Like large animals, rectal palpation in small ruminants is not possible due to small sized pelvis (Goel and Kharche, 2009). By using real time B-mode ultrasonography pregnancy can be detected 30 days post insemination using a modified human probe to make it suitable for goats (Goel *et al.*, 2009).

Ovulation can be detected by the acute disappearance of a large follicle (i.e. black dot converted to a white area) and subsequent formation of corpus luteum.

Ultrasonography has provided an accessible method of monitoring superovulation.

The number of developing follicles can be counted and followed to see the number of corpora lutea formed. The superovulated animals, if examined before flushing can give an evaluation of the superovulatory response and the identification of any unovulated follicles. For animals that have not been superovulated, the flushing procedure may be appropriately abandoned to save time and money. Kumar *et al.* (1997) studied the ovarian follicles of buffalo during superovulation using ultrasonography. It was found that pool of preantral follicles is responsible for the number of large follicles destined to ovulate. Sharma *et al.* (2001, 2008) described the use of ultrasonography in the diagnosis of liver diseases. The merits of ultrasound are its ability to characterize internal parenchymal architecture, cost effectiveness and no radiation hazards.

The greater use of ultrasonography in future may ensure higher success rate of embryo transfer studies. Another application using ultrasonography is the collection of oocytes from live animals for use in IVM/IVF studies. The use of ultrasonography is being tried to perfect the procedure of sex determination in farm animals.

A-mode ultrasound : (amplitude modulation) that in which on the cathode-ray tube (CRT) display one axis represents the time required for the return of the echo and the other corresponds to the strength of the echo, as in echoencephalography. Recently ultrasonography has replaced the A-mode ultrasound non imaging system used in the past (Goel and Agrawal, 1992).

B-mode ultrasound : (brightness modulation) that in which the position of a spot on the CRT display corresponds to the time elapsed (and thus to the position of the echogenic surface) and the brightness of the spot to the strength of the echo; movement of the transducer produces a sweep of the ultrasound beam and a tomographic scan of a cross-section of the body. Ultrasonography is not, however, of much value in examination of the lungs because ultrasound waves do not pass through structures that contain air.

Presently, real time : **B**-mode ultrasonography is used for the determination of early pregnancy in livestock (Goel *et al.,* 2009). In case of goats, B-mode real time transrectal ultrasonography using frequency of 5 to 7 mHz can be effectively used for early pregnancy diagnosis (4 week of gestation) and to study early embryonic and foetal development in goats.

A particularly important use of ultrasonography is in the field of obstetrics and gynecology. It is a fast, relatively safe, and reliable technique for diagnosing pregnancy, and for detecting some typical fetal anomalies in both human and domestic animals. Ultrasonic imaging in animal reproduction started in the 1970's and in 1980 the first paper, by Palmer and co-workers on ultrasonic scanning of the uterus and ovaries in the mare was published. The use of ultrasonography has been tried to perfect the procedure of sex determination in farm animals (Muller and Wittkowski, 1986; Wideman *et al.,* 1989). The use of ultrasonic imaging technique for diagnostic purposes in veterinary medicine and animal science has lagged far behind their use in human medicine.

Doppler ultrasound : That in which measurement and a visual record are made of the shift in frequency of a continuous ultrasonic wave proportional to the blood-flow velocity in underlying vessels; used in diagnosis of extracranial occlusive vascular disease.

Endoscopic ultrasound : A high resolution ultrasound transducer, mounted on a flexible endoscope, can be used to gain images from within a hollow organ, such as the gastrointestinal tract. This overcomes some of the problems of ingesta and fecal material cause in other methods of ultrasound examination.

Gray-scale ultrasound : B-mode ultrasonography is one in which the strength of echoes is indicated by a proportional brightness of the displayed dots.

M-mode ultrasound : (motion mode) a type of B-mode ultrasonography in which spots on the CRT display produce a tracing of the motion of echogenic objects. It is used in echocardiography.

Real-time ultrasound : B-mode ultrasonography using an array of detectors so that scans can be made electronically at a rate of 30 frames a second, thus giving a true display of motion, such as that of the heart.

Computer-assisted image analysis and ultrasonographic assessment of ovarian follicular development are some of the extensions of the technological advances in diagnostic ultrasonography.

Risks : Because ultrasonography uses high frequency sound waves, and not X-rays or other forms of radiation, there are very few risks associated with its use. Sound waves are either reflected back to the transducer, or the tissues of the body absorb them and they dissipate as heat. There may be a slight increase in heat in the body as a result, but no negative effects of this heat have been documented.

The main disadvantages of the use of ultrasonography are related to cost and time involved with the use of this technique. Ultrasound machines are expensive and it takes more time to perform a pregnancy diagnosis with an ultrasound machine than by rectal palpation.

In conclusion, ultrasound has become very important for non-invasive observation of soft structures in real-time. It is an interactive diagnostic technique, which requires much experience to find the right view for proper diagnosis.

Ultrasound-guided Transvaginal Oocyte Pick-up (OPU)

Ultrasound-guided transvaginal oocyte aspiration provides an excellent non-invasive method for obtaining oocytes. This technique has been in use in cattle and buffaloes. The recovered oocytes can then be used in *in vitro* fertilisation or other programmes or used for other purposes. *In vitro* fertilisation (IVF) programmes are well established in human medicine and a key part of the process is ultrasound-guided transvaginal oocyte aspiration and this technique has superseded the more invasive laparoscopic techniques. In animals, oocytes have traditionally been obtained from follicle aspiration of slaughterhouse ovaries or via a flank laparotomy technique. Slaughterhouse material has the obvious disadvantage of lack of repeatability and there is often considerable delay in the time between oocyte collection and placement in culture medium. Surgical laparotomy techniques have obvious disadvantages in terms of ease and repeatability.

Pieterse and his co-workers at Utrecht University (Pieterse *et al.*, 1988), using cattle, were the first to describe the technique of oocyte aspiration during transvaginal ultrasound ovarian scanning in domestic animals. When heifers reach puberty at 11 to 12 months of age, their oocytes may be retrieved weekly or even twice a week for embryo production and embryo transfer. There is even the possibility of applying this technology to juveniles. In this way, high value female calves can be used for breeding long before they reach their normal breeding age. Transvaginal ovum pick-up (OPU) offers several advantages over standard embryo transfer procedures ie laparoscopic OPU (L-OPU) method eg the greater changes in vacuum pressure during L-OPU vs U-OPU might be responsible for the difference in oocyte quality (Santl *et al.*, 2009). Generally the ultrasonographic method is less traumatic.

Some form of casing around the transducer is needed in order to guide the needle into the follicle to be aspirated. In the initial studies of Pieterse *et al.* (1988; 1991a,b) a 7.5 MHz sector transducer was inserted into a 50 cm long autoclavable stainless steel holder with a handle at one end to facilitate manipulation. The casing had a needle guide channel running the length of the casing. On the ultrasound image as displayed on the monitor there is a needle/biopsy guideline which corresponds to the location and direction of the aspiration needle, making it possible to predict the direction of passage of the needle on the monitor screen. A variety of needles have been used of varying lengths and thickness' and either specially designed ovum pick-up needles or ordinary luer-lock disposable needles. Suction is necessary to aspirate the follicles and recover the fluid containing the oocyte. Special technique is required to accurately puncture the follicle and flushing of the oocytes. Recovery rates using this technique have been variable due to several reasons. Several OPU devices that are suitable and economical for routine use in oocyte retrieval have been developed (Manik *et al.*, 2003).

Method for producing IVP embryos must take due account of the risks of disease transmission.

Embryo Cloning or Nuclear Transfer

Cloning may be defined as asexual reproduction, or as the creation of genetically identical individuals. Proliferation of identical fragments of DNA (molecular cloning) is also referred to as cloning, but is clearly different from somatic cloning of organisms. There are four different artificial methods to produce clones in mammals.

Blastomere Separation

The simplest of the methods of embryo cloning involves manipulating early embryos just a few days after fertilization. Individual embryonic cells (called blastomeres) upto the four-cell stage are undifferentiated and each cell has the potential to develop into an offspring. Thus each cell is 'totipotent'. The separated cells simply continue to divide according to their original developmental programme to form a new organism.

Embryo Splitting

Slightly older embryos at the morula or blastocyst stages may be cut into two equal halves by using a micromanipulator and a microsurgical knife, before transfer to a surrogate female.The genetically identical animals produced by these methods are limited. This process seems to mimic the natural process of production of monozygotic twins.

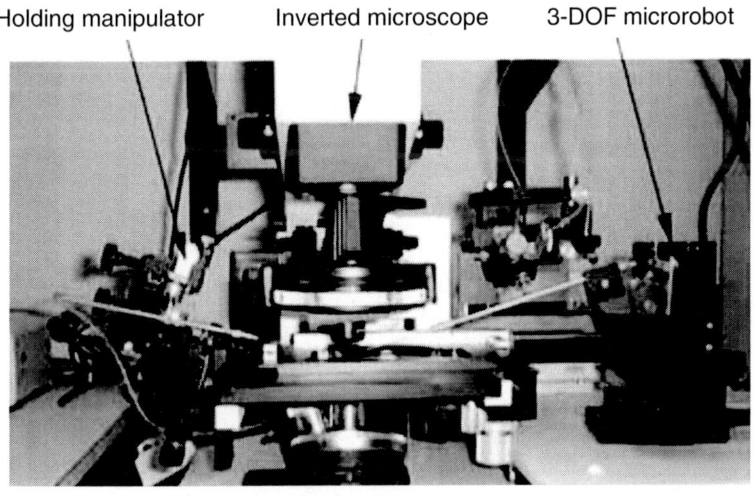

Micromanipulator assembly for manipulating embryos
(Some of important companies manufacturing micromanipulator include Zeiss, Wild, Nikkon etc)

Nuclear Transfer Cloning (NT)

Embryonic Stem Cell Complementations

Derivation of cloned animals from pluripotent embryonic stem cells is possible if complemented with suitable helper cells. Embryonic stem cells are typically derived from the inner cell mass of blastocyst-stage embryos and can form all of the tissues of the foetus.The potential has been demonstrated in mouse but yet to in livestock (Wells, 2006).

Fertilization of mammalian egg is followed by successive cell divisions and progressive differentiation, first into the early embryo and subsequently into all of the cells types that make up the adult animal. Transfer of a single nucleus to an enucleated unfertilized egg and subsequent development provides a unique opportunity to produce clones by nuclear transfer. In February 1997, cloning became the focus of media attention, when Ian Wilmut and his colleagues at the Roslin Institute announced the successful cloning of a sheep, named Dolly, from the mammary glands of an adult female. Wilmut *et al.* (1997) reported the production of a sheep, "Dolly", by the transplantation of a mammary cell nucleus from an adult ewe to the enucleated oocyte of another sheep.

Cloning by nuclear transfer involves the removal of the nucleus from one cell and its placement in an unfertilized egg cell whose nucleus has either been deactivated or removed. The enucleate ooplasm with its mitochondria and other organelles in this achievement served as sufficient medium in which to express the genetic heritage of a nucleus derived from a growth-arrested, differentiated somatic cell. The oocyte or unfertilized egg serves as a recipient. The DNA comes from the donor cells of the organism being cloned. The donor cell can come from a variety of sources, including early embryos, fetuses or adults (eg in 'Dolly" sheep was cloned from adult mammary gland cell).

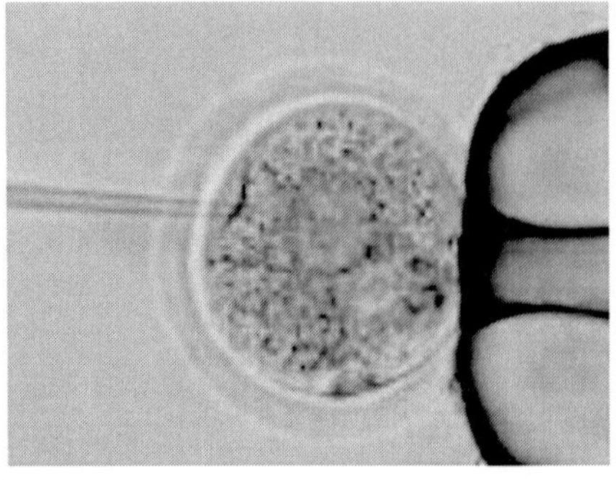

Injection of foreign DNA into oocyte. The holding pipette keeps the oocyte in place while injecting pipette with the help of DNA injector introduces foreign DNA into cytoplasm

This process could also be viewed as the production of a nucleus/ cytoplasm chimera. The introduction of a diploid nucleus into an enucleate egg, can be aimed to produce only one individual.

This method has become very popular because of the potential to produce extremely large number of clones that are genomic copies of selected individuals.

On June 6, 2009, Scientists at National Dairy Research Institute, Karnal announced the India's - and the world's - first cloned water buffalo calf (Singla *et al.*, Personal communication) using a hand guided method. Hand-guided cloning needs a micro manipulator.

The hand guided method used by NDRI is essentially as follows (Singla *et al.*, 2009 http://www.business-standard.com/india/news/cloned-milkcereals/367718/).

Dolly : the cloned sheep

Step 1: Two eggs are taken and the "egg shells" are removed using enzymes

Step 2: The nucleus material "yolk" of the eggs is removed. To do that, enzymes are used to push it to one side. The resulting "bulge" is cut with a finely sharpened micro blade

Step 3: The nucleus of the chosen cell (taken from the animal to be cloned) is extracted. Skin cells are chosen, because they are the least specialised

Step 4: The nucleus is sandwiched between the two eggs and a low-voltage electric current is applied. This results in all the material fusing

Step 5: The resulting egg with the foreign nucleus is incubated. If it becomes a healthy, growing embryo, it is chosen. The embryo, before it becomes 100 cells in size, is further manually divided to form multiple identical embryos (identical twins)

Step 6: Four identical embryos are implanted in the uterus of each buffalo chosen as a surrogate mother. The buffalos have had their fertility monitored and are treated to ensure they don't reject the embryos.

Step 7: If a pregnancy results in the implanted buffalos, then you have a good chance of a cloned calf being born.

Adult sheep, goats, cattle, mice, pigs (Prather *et al.*, 1989), cats and rabbits have been cloned using somatic cell nuclear transfer (Edwards *et al.*, 2003).

Limitations and Future Prospects of Cloning

The cloning of Dolly made it apparent to many that the techniques used to produce her could someday be used to clone human beings. This stirred a lot of controversy because of its ethical implications. First animal among the domestic species cloned by somatic cell nuclear transfer was a small ruminant ie Dolly sheep. A major limitation of cloning procedures is the extreme inefficiency for producing live offspring. Dolly was the result of a batch of 277 embryos reconstructed by transfer of nuclei of cultured adult mammary epithelial cells, a success rate of 0.4 % (Wilmut *et al.*, 1997). Similar inefficiencies for cloning adult animals of other species have been described by others. Death of cloned embryos and fetuses occur throughout pregnancy. Moreover, a high proportion of cloned offspring are generally larger than normal (large offspring) and die soon after birth. Many factors related to cloning procedures and culture environment contribute to the death of clones, both in the embryonic and fetal periods as well as during neonatal life. The high rate of early embryonic mortality occurring during the first trimester of cloned pregnancies may be due in part to inappropriate expression of trophoblast major histo-compatibility complex (MHC) class I antigens. Extreme inefficiencies of this magnitude, along with the fact that death of the surrogate may occur, continue to raise great concerns with cloning humans.

The major future use of nuclear transfer in small ruminants is in the area of transgenic animal production for protein harvesting. Adult sheep, goats, cattle, mice, pigs, cats and rabbits have been cloned using somatic cell nuclear transfer (Edwards *et al.,* 2003).

The transfer of totipotent nuclei in ennucleated oocytes theoretically allows for the production of large numbers of identical twins or clones. This opened prospective to affect genetic response in a variety of ways including selection intensity, selection accuracy and generation interval. Initially, the sources of totpotent nuclei were blastomeres. Despite the potential use of first as well as higher order generation blastocysts as nuclei donors, the size of the clones has remained very small. The recent generation of totipotent embryonic stems 'ES' - like cells in sheep which will likely be followed by similar developments in other species, might lead to a considerable increase in the efficiency of embryo cloning. Application of, cloning in the propagation of endangered species and conservation of their gene pools is also a possibility in future. Domestic animal cloning also has potential applications in biomedicine.

Sexing of Embryos and Separation of X and Y Chromosomes Bearing Sperms

Nettie Maria Stevens (1861-1912) discovered that chromosomes specifically, X and Y chromosomes found in sperm-determine sex. The Bryn Mawr College-based geneticist published her discovery in 1905, at approximately the same time Edmund Beecher Wilson, a Columbia University biologist, made the same observation. Stevens was born in Cavendish. The X and Y chromosomes are the sex chromosomes for mammals, including humans. Not only are the X and Y sex chromosomes in mammals physically distinctive, with the Y being smaller, the Y chromosome is exceptionally peculiar. The X chromosome contains considerably more genes than the Y, which has its functionality essentially limited to traits associated with being male. It is the Y chromosome that carries the major masculinity determining gene (SRY, for sex-determining region Y), which dictates maleness. In a mating pair, if the paternal partner contributes a normal Y chromosome, male gonadal tissues (testes) develop in the offspring. Only males have the potential to transmit a Y chromosome to the next generation, and thus the father's contribution is decisive regarding an offspring's sex.

Several methods have been tried in the past to separate the X and Y bearing spermatozoa with questionable success.

One of the methods used was to carry out embryo sexing by detection of H-Y antigen. This technique was about 85% accurate. This involved the immunological detection of male specific antigen referred to as H-Y antigen.

In another method, measurement of X-linked enzyme was carried out. It was seen that D-linked enzyme activity was twice in case of female embryos than male embryos. This method is also not highly accurate.

The Karyotyping technique involves the biopsy of a small number of cells of an embryo and their metaphase chromosomes are examined microscopically for the detection of two X- chromosomes or one Y chromosome.

PCR in Sex Determination

In this technique, DNA from the animals is amplified by the PCR using primer systems and sex specific products are identified by restriction fragment length polymorphism (RFLP) analysis. Considerable success has been achieved in getting correct sex determination using this technique. Embryo sexing can also be achieved by micro biopsy and sex determination using PCR amplified V-specific sequences. This approach, however, is economically justified only in very exceptional circumstances. This might become the method of choice to generate embryos of a desired sex.

In the DNA probing method, biopsy of embryo is performed to determine the sex by localization of DNA sequences present on the Y chromosome.

The most satisfactory, accurate, sensitive and specific rapid and reproducible method of sexing of embryos is polymerase chain reaction. The application is to amplify Y-chrmosome specific (male specific) DNA sequences, directed by primers with a thermostable DNA polymerase from a small amount of DNA of one or more blastomere biopsied from the embryo.

The most recent method, which shows some promise for the future is the use of cell sorting machine to separate the X and Y bearing spermatozoa based on minute differences in the weights in DNA of sperms of two types. Larry Johnson of University of California was able to achieve some success using this technique of rabbit spermatozoa (Johnson and Clarke, 1988). The technology, by Lawrence A. Johnson of the U.S. Department of Agriculture in Beltsville, Md., exploits the difference in amounts of DNA in X and Y chromosomes. Sperm bearing the X, or female, chromosome have more DNA than sperm carrying the Y, or male, chromosome. An embryo resulting from the merger of an egg, which always carries an X chromosome, and a sperm carrying an X chromosome will have two Xs and, therefore, develop into a female. An egg fertilized by a sperm carrying a Y chromosome becomes a male.

Sex determination of livestock is performed to increase the profitability as well as productivity of the livestock production system. In the livestock industry, it is desirable to control the sex ratio of animals. The dairy animal improvement benefits from the birth of female offspring. For meat production, male offsprings are generally advantageous because male animals tend to grow faster and have better feed efficiencies. Sexing of pre-implantation embryos from livestock species is of a great importance in transgenic because it provides a way to choose the desired sex.

The most promising technique to sex embryos obtained by flushing or by IVF techniques; at present available is based on the polymerase chain reaction (PCR). PCR is fast, and because it amplifies the DNA, only tiny initial amounts of DNA are needed. This technique is based on the amplification and identification of Y chromosome sequences. Pomp *et al.* (1995) reported success with this technique in a variety of farm animals and the technique used by them took approximately 6 h from receipt of embryos to results.

Rao and Totey (1992) studied the sex determination in sheep and goats using bovine Y chromosome specific primers via polymerase chain reaction. In an attempt to delineate factors contributing to the success of cleavage and develoment of transplanted embryos, it was observed that insulin like growth factor-1 (IGF-1) enhanced these for buffalo embryos. Temporal and spatial expression of IGF-1 is being examined to explain the mechanisms involved.

Several approaches have been established for genetic sex typing, such as karyotyping, H-Y antigen detection, X-linked enzymatic determination and the polymerase chain reaction. PCR-based embryo sexing is becoming prevalent as compared to other methods as it requires less time, easy to perform and better accuracy as compared to other methods. PCR application allows much faster and more precise classification of embryos by sex than tradition karyotyping or enzymatic procedures. Various methods have been used for determining the sex of mammalian embryos. Successful sex-determining methods based on the amplification of Y chromosome specific sequences have been developed in livestock. Sexing using PCR involves the co amplification of the Y-chromosomal sequence and an autosomal sequence, which acts as control for testing biopsy material. PCR is designed to yield different fragment sizes for the Y-chromosomal and the autosomal products, which are separated by electrophoresis using ethidium bromide stained agarose gel.

Transgenic Animal Production

The newer tools and technologies which have been developed by the scientists allow un-precedented control and power to manipulate the genetic material of organisms. These technologies/tools have formed the subject of the new discipline called biotechnology. These tools and technologies, especially the advent of recombinant DNA has sparked the imaginations of the scientists and laymen alike. One of these technolgies which are responsible to bring about recent revolution in this field is genetic engineering enabling us to transfer genes from one organism to another and bioprocess technologies to produce large quantities of biological drugs. Transgenic technology is a powerful tool for improving the production characteristics of livestock. The introduction of specific genes into the genome of farm animals and its stable incorporation into the germ line has been major technological advance in agriculture.

There are a number of methods that can be used to produce transgenic animals. However, the primary method to date has been the microinjection of genes into the pronuclei of zygotes. Transgenic technology provides a method to rapidly introduce "new" genes into cattle, swine, sheep, and goats without crossbreeding, production of foreign genes in convenient procaryotic cells and the large scale production of gene products. Practical applications of transgenesis in livestock production include enhanced prolificacy and reproductive performance, increased feed utilization and growth rate, improved carcass composition, improved milk production and/or composition, and increased.

These protein products could have application as therapeutics, diagnostics, restriction enzymes or industrial enzymes. At present there are several biotechnolgy derived therapeutics approved for human use and the market value of such products run in billions of US $. The second technology could be use of transgenic eurkaryotic animal's especially small or large ruminants and harvesting the milk containing modified products.

The foundation for the production of transgenic animals was provided by the pioneering experiments of Brackett *et al.* (1971) using sperm mediated gene transfer in rabbits. Production of eukaryotic animals as transgenic became a real possibility, when from 1980 onwards, several research papers were published, which demonstrated that DNA micro injections could produce transgenic mice. Transgenic mice have provided powerful tools for the study of regulation of gene expression in physiological functions. Gene therapy was though to be viable technology which could be made availabel by refinements in gene injection protocols. Similarly transgenic animals have been thought of as live pharmaceutical units for the production of biologicals. As it is known that the conventional chemical methods have limitations in synthesis of large bio-molecules like proteins etc. The proteins are useful for many industrial and therapeutic purposes. Purification of such proteins from live animal sources was a very cumbersome and uneconomic process. By use of biotechnology, especially transgenic animals these proteins are likely to be available in sufficient quantities economically. Rare human proteins which cannot be produced by microorganisms because of requirements for post translational modification can be secreted into milk of transgenic animals. Large amount of biologically active proteins can be obtained from transgenic farm animals using this system.

The overall goal in making a transgenic animal is the stable introduction of a desired DNA sequence into the germ line of the host animal that can be transmitted to offspring in a Mendelian fashion. By incorporating new or modified genes at the genetic level, the characteristics of the animal can be specifically changed. Transgenic animals are generated for a variety of purposes. They can be used as basic research models, specialized non-agricultural purposes (such as pharmaceutical production or xenotransplantation) and also to enhance animal production traits and products. For many applications, large animals, e.g., livestock such as pigs, cows, sheep, and goats, are of interest. Producing transgenic livestock is not as efficient as mice and is an expensive and time consuming process. Accordingly, there is much interest in developing methods that increase the efficiency and specificity of the transgenic process in non-murine large animals.

Transgenic animals are generally produced by one of three main methods: 1) the pronuclear microinjection of fertilized one-cell embryos followed by reimplantation into surrogate mothers; 2) the genetic manipulation of embryonic stem (ES) cells followed by introduction of modified ES cells into developing embryos; and 3) by the genetic manipulation of somatic primary cells followed by nuclear transfer into a recipient oocyte. The standard and most established method of producing transgenic animals such as mice, rabbits, pigs, goats, or cows generally rely on the microinjection of DNA encoding a transgene into the pronucleus of fertilized zygotes. However, this method currently has several unavoidable shortcomings.

Pronuclear microinjection methods generally result in the random integration of transgenes in the chromosome of the zygote. If the DNA construct is integrated into an inactive region of chromosome, it is unlikely to be expressed. As a consequence, it is necessary to generate several founders and carry out extensive characterizations on them all in order to identify a line of animals that will stably express the transgene at appropriate levels. DNA construct design is also crucial when using pronuclear microinjection. Promoter and regulatory elements must be present in the DNA fragment injected in order to dictate when and where the transgene will be expressed. The optimization of transgene design is time consuming and labor intensive. When using pronuclear microinjection, only gene additions at a random location are feasible until recently. The complete removal, mutation or replacement of endogenous genes is not possible. Furthermore, the efficiency with which transgenic animals are generated with this technique is quite low.

Several of the problems discussed above can be circumvented by introducing DNA via the transfection of ES cells or by cloning, both of which allow for the targeted insertion of DNA into cells in culture. The important feature of these methods for the production of transgenic animals is that both ES cells and any donor cell (i.e., the differentiated somatic cell) to be used in nuclear transfer can be grown in culture and genetically modified with a desired transgene. The modified cells can then be selected, characterized prior to being used to generate transgenic animals. The potential advantages of these methods offer over pronuclear microinjection include the ability to do gene targeting, thereby allowing for the creation of knockouts and enabling the modification of endogenous genes. Also, with cloning, all animals born will be germ line transgenic. However, identifying the homologous recombinants in a large population of non-homologous random integrants often proves to be the rate-limiting step for creating homologously modified mammalian cell lines. This severely limits the ability to manipulate target genes systematically.

These strategies are labor-intensive, time-consuming, and ultimately limit homologous recombination genetic engineering of mammalian cells for commercial applications. Other disadvantages include the fact that currently, among mammals, ES cells are available only for mice. While nuclear transfer allows for targeted modifications in livestock species, it is not supportive with all cell types, requires specialized techniques and conditions, is hard to maintain pregnancies and is associated with large offspring syndrome. Moreover, the process is also very inefficient. The efficiency and frequency with which transgenic animals are generated with these methods are in the same range as those of the more established and simpler method of pronuclear microinjection. Presently, nuclear transfer efficiency in sheep is around 0.04 to 1.7% live born animals from reconstructed embryos, which is similar to standard pronuclear microinjection transgenic rates of approximately 1%.

Thus, there is a need in the art for methods of increasing the efficiency of generating transgenic animals, particularly livestock.

Production of transgenic animal is a difficult and cumbersome process at present. Injection of cloned DNA into the pronuclei of fertilized embryos is currently the best way to produce transgenic animals. Exogenous DNA can integrated into the host genome and passed from parent to progeny. Microinjection of foreign DNA involves use of micro tools, micro pipettes, micro manipulators and other DNA injectors. The efficiency of producing transgenic livestock is much low as compared to that of producing transgenic mice.

Growth related applications to livestock species using growth hormone genes or growth factor genes have been disappointing. There were many undesirable side effects noted in the transgenic animals. More sophisticated regulatory systems are needed to control expression of transgenes in the transgenic animals. Turning on and off transgene expression in response to endogenous or exogenous signals may allow for desired positive effects while circumventing potentially harmful effects.

Not much work has been done on transgenic production in India. Two research groups, one from Madurai Kamraj University and another from National Institute of Immunology have reported the production of transgenic fish. A research initiative by DBT at Bose Institute Kolkata did not result in transgenic production so far. However, the future holds immense possibilities for research in this area. Major hurdles in such type of research are high cost of biologicals and equipments. A microforge, pippete puller, micro manipulator, DNA injector, DNA transilluminator, CO_2 incubator, DNA sequencer, thermal cycler are some of the requirments for setting up a transgenic lab. Some of the groups in USA have achieved considerable success in this area and are producing transgenic animals as a matter of routine. Some biotech labs in the private sector in the west have already harvested the gains of transgenic research on a commercial scale. All we need to put our best efforts in this direction and provide resources for such research.

Since 1980, when the first successful gene-transfer experiment using DNA microinjection on a mouse was reported (Gordon *et al.*, 1980), a method has been established that allows the transfer of a single isolated gene. By using molecular biology methods already established in the 1970s, genes that are of interest for breeding purposes can be isolated, sequenced, recombined with regulatory elements and tested *in vivo* and *in vitro* for their functional ability. The implementation of this basic research in animal breeding programmes is only just beginning and it is expected that drastic changes will emerge in the future. Gordon *et al.* (1980) first product transgenic mice by microinjection of plasmid DNA. Since transgenic mice were first produced by DNA microinjection, the potential has been great for genetic engineering of animals through introduction of extra copies of cloned genes into the animal genome. However, the overall efficiency of producing transgenic live stock species is low compared to that of transgenic mice. One promising practical application of gene transfer technology to livestock species is the production of pharmaceuticals. While conventional breading strategies as well as marker assisted selections are restricted to the exploitation of genetic variation pre-existing within the species, in breed of interest, transgenics has opened the exciting possibility to exploit variants across species barriers or even create de novo. The traditional approach to produce transgenic animals was micro injection of DNA constructs in the male pronucleus of one celled stage embryos. Recently (Campbell *et al.*, 1996) has given a new technique i.e. by genetic targetting of totipotent cells in culture, followed by either nuclear transfer in enucleated oocytes or micro injection into blastocysts. Until now suitable totipotent livestock cell lines were not available. The description of ovine 'ES' cells and their use to produce chimeric sheep, however, demonstrates that gene targetting methods might become available for animal 'production in near future (Campbell *et al.*, 1996).The generation of rat spermatogenesis in mouse testes suggests that spermatogonial stem cells of many species could be transplanted and opens the possibility of xenogeneic spermatogenesis for other species. Alok *et al.* (1995) from Nil, New Delhi reported the transfer of Human growth hormone gene into Indian Major Carp (Labeo rohita). Two of the fish developed from microinjected embryos were transgenic and were found to carry 1 or 2 copies of the human growth hormone gene in their genome.

For gene transfer in mammals, three different techniques are possible:

- Microinjection of DNA into the pronuclei of zygotes,
- DNA transfer using retroviral vectors,
- Production of transgenic chimeras by injecting genetically transformed totipotent stem cells into embryos.

Gene Transfer Through Direct DNA Microinjection

Until now the microinjection of DNA into the pronuclei of zygotes is the only method used for gene transfer in domestic animals. The injection of a DNA solution into the pronuclei of fertilized eggs is the most common method for making transgenic animals.The procedure consists of the following phases:

- Cloning and recombination of a suitable gene construct,
- Preparation of donors,
- Recovery of zygotes (fertilized oocytes),
- Visualization of pronuclei (necessary in some domestic animal species),

- Preparation of the DNA solution to be injected,
- Microinjection of DNA solution into the pronuclei of the zygotes, Injection is done at the stage of development when mammalian ova have two pronuclei, one from each gamete, which will later fuse to form the diploid nucleus.
- Transfer of injected zygotes into the oviducts of synchronized recipients,
- Investigation of newly born animals to ascertain whether they have integrated the gene construct (Dot - Southern - Blot).

The equipment needed for microinjection includes an invert microscope, two micromanipulators and injection equipment. Additional necessary items are an injection chamber and holding and injection pipettes. The injection pipette, with an outer diameter of 1 to 2 m m, is filled with DNA solution. During injection, the zygote is held with the holding pipette. The injection pipette is introduced into the pronucleus, passing through the zone pellucida, the cell membrane and the nucleus membrane. About one to two picolitres of the DNA solution are injected into the pronucleus, increasing its volume.

These injected zygotes are transferred, after brief *in vitro* culture, into the oviducts of synchronized recipient animals. Recipient animals are synchronized with the donors by means of hormonal treatment. After the birth of offspring from gene injection, high molecular DNA is isolated from tissue (blood or cells can be conveniently taken from the tail), to confirm successful integration. The integration of the injected DNA and the number of integrated copies can be determined after further processing with **Southern Blot or Dot Blot hybridization**. Integration sites in the chromosomes can be proved through hybridization of metaphase chromosomes, using the injected gene as a probe.

Transgenic animals are raised and mated. Offspring from these matings are tested to discover whether the transgene has been passed on. In mating hemizygous-transgenic F1 inter se attempts are made to produce homozygous transgenic animals.

Gene Transfer with Retroviral-Vector Assistance

Retroviruses have an RNA genome, which is transcribed through the virus' own reverse transcriptase into DNA in infected cells and subsequently integrated into the genome of the cell. Integration occurs accidentally. However, only one copy can generally be found in the provirus. Based on this cycle retroviral vectors can be used as vehicles for gene transfer.

Gene Transfer through Production of Germ-line Chimeras

The procedure to utilize totipotent transformed stem cells to transfer recombined gene constructs into the germ line is receiving increasing attention, especially in respect of the mouse. Totipotent stem cells are isolated from *in vitro* cultured blastocysts. Through aggregation with early embryonic division stages or through injection of these cells into blastocysts, chimeras can be produced. Up to 30 percent of the chimeras so produced are germ-line chimeras containing the genotype of the cell line.

In recent years, several application possibilities for gene transfer in domestic animals have been discussed. Until now it has been possible to influence only traits that are based on a single gene or on a limited number of genes. There are only a very limited number of traits of interest to breeders, which are based on a single gene. The difference between qualitative and quantitative traits is not always fully clear. For future research activities in the field of gene transfer, the following three subjects should be given particular attention.

- Investigations to improve the effectiveness of gene transfer methods. While trying to optimize the technique of gene transfer through microinjection, attempts should also be made to establish and to optimize the other techniques in gene transfer.
- Isolation and characterization of genes of interest for breeding purposes. The results of basic research in molecular biology are an excellent foundation for practical work on the genome of domestic animals.
- Isolation and characterization of suitable regulation elements. The provision of suitable promoters and enhancers is even more difficult than the search for structure genes. For convenient use in breeding programmes, it is of the utmost importance to produce gene constructs that allow the transferred constructs to be expressed in the right tissue at the right time, as well as in the right amount.

Parthenogenetic Embryos: An Emerging Tool in Reproductive Biotechnology for Multiplication of Superior Germ Plasm

Biologically, Parthenogenesis (Gr.,= virgin birth) is a form of reproduction in which the ovum develops into a new individual without fertilization. This process occurs naturally in aphids, daphnia and some other invertebrates including many plants. Vertebrates like Komodo dragons and sharks have recently been added to the list of vertebrates along with several genera of fish, amphibians and reptiles. The phenomenon of parthenogenesis was discovered in the 18th century by Charles Bonnet. In 1900, Jacques Loeb accomplished the first clear case of artificial parthenogenesis, when he pricked unfertilized frog eggs with a needle and found that in some cases normal embryonic development ensued. Numerous mechanical and chemical agents have been used to stimulate unfertilized eggs. In 1936, Gregory Pincus induced parthenogenesis in mammalian (rabbit) eggs by temperature change and chemical agents. Production of parthenogenetic embryos has been successful in different animals including human beings. The developmental capacity of parthenogenetic embryos has been extensively investigated with respect to pre-implantation and post-implantation development not only in mice but also in domestic animal such as pig, sheep and bovine. In bovines parthenotes had established pregnancies up to 67 days. In mammals, genomes from both parents are generally needed to make viable offspring. But changing the expression of 'imprinted' genes can render the father's contribution dispensable. Sexual reproduction in animals ensures that each individual normally inherits one set of genes from each parent. But viable offspring that have only maternal genes none from the father can be produced through parthenogenetic reproduction in plants and most groups of animals. Recently, Kono *et al.* (2004) produced the first parthenogenetic mice. Partheno-genetically oocytes could also be used to establish embryonic stem cell line. On August 2, 2007 South Korean Scientist, Hwang Woo-Suk, produced the human embryos through parthenogenesis, which opened vista for creating stem cells that are genetically matched to particular woman for the treatment of degenerative diseases. In mouse, germ line competent as well as histocompatible parthenogenetic ES cell lines have been derived. Suppression of meiosis and activation of oocyte to develop without involving sperm would be the ideal approach to produce offspring. This could lead to the production of superior livestock without the need for male counterpart in sexual reproduction.

Thus animal cloning through parthenogenetic cell will help in multiplying the few superior germplasm throughout the country, which will help in enhancing the production.

The field of *in vitro* maturation and *in vitro* fertilization of bovine follicular oocytes has progressed enormously since 1980. Zygotes, embryos, and calves derived from the *in vitro* maturation and fertilization of immature bovine follicular oocytes have been produced both for commercial and research purposes; however, reports show only a modest percentage of success rate. The reasons for the low pregnancy rate and the low number of live calves are not precisely known. One reason might be that parthenogenetic activation of oocytes might occur in such systems. Oocytes matured *in vivo* or *in vitro* can undergo parthenogenetic activation spontaneously *in vivo* or in vitro. The parthenogenetic embryos are uni-parental embryos. There are two types of uni parental embryos i.e. andro-genetic embryos and gyno genetic/partheno-genetic embryos. Andro genetic embryo has two sets of equivalently imprinted male chromosomes.

Androgenetic Embryos: There are two methods of androgenetic embryos construction i.e. *In vitro* fertilization of enucleated oocytes and traditional method. The traditional method consists of replacement of female pro-nucleus with a male pro-nucleus in the zygote. It can be performed conveniently by Piezo-intra cytoplasmic sperm injection (ICSI). These methods has been applied successfully only in mice. It has been demonstrated that androgenetic embryos can develop not only up to the 6 to 8 somatic stage as in the case of the pronuclear transfer but also up to 25 somatic stages. Modification of the method of androgenetic embryo production from enucleation of fertilized oocytes to fertilization of enucleated oocyte has increased the development of mouse haploid androgenetic to the morula (30%) and blastocyst stage (11.5%). However the development of haploids remained significantly lower as compared to diploid androgenotes (43% and 54%, respectively). The application of the second method in the bovine is limited because the size difference between the two pronuclei is not as obvious as in mice. Therefore, it is impossible to be certain of the origin of the pronuclei. In addition, bovine zygotes have an opaque cytoplasm that requires a centrifugation step to visualize the pronuclei. For this reason, the only possibility for correctly identifying the male pro-nucleus in bovine is to fertilize enucleated oocytes by intra cytoplasmic sperm injection (ICSI) or by *in vitro* fertilization (IVF).

A form of asexual reproduction related to parthenogenesis is gynogenesis, where offsprings are produced by the same mechanism as in parthenogenesis, but with the requirement that the egg be stimulated by the presence of sperm in order to develop. However, the sperm cell does not contribute any genetic material to the offspring. Since gynogenetic species are all female, activation of their eggs requires mating with males of a closely related species for the stimulus needed. Some Salamanders of the genus Ambystoma are gynogenetic and appear to have been so for over a million years. It is believed that the success of those Salamanders may be due to a rare fertilization of eggs by a male, introducing new material to the gene pool, which may result from perhaps, only one mating out of a million.

In bovine oocytes, activation by Ca^{2+} ionophore and ethanol results in a wide range of parthenogenetic development from cleavage up to blastocyst formation. It has been suggested that activation is mediated by an increase in Ca^{2+} ionophore that in turn leads to the exit from the second meiotic block via the inactivation of the cytostatic factor and consequently the mitotic promoting factor. Parthenogenesis has been induced in bovine oocytes by exposure to a $Ca^{2}+$ ionophore followed by cycloheximide or 6-dimethylaminopurine (6-DMAP), resulting in the resumption of embryonic cell cycles and a high percentage of blastocyst formation. The effects of 6-DMAP on chromatin and microtubule configurations during meiosis are mediated by the inhibition of protein kinases and, consequently, protein phosphorylation. Dephosphorylations induced by 6-DMAP inactivate c-mos and mitogen-activated protein (MAP) kinase, as well as a range of downstream unidentified kinases leading to the rapid formation of a nuclear envelope after oocyte activation, but with consequences for the normal progression of post fertilization events. Treatment with 6-DMAP after bovine oocyte activation induces pronuclear formation and drives the parthenote into interphase of the first mitotic cell cycle, presumably as a uniform diploid. However, examination of chromosomal complements in bovine parthenotes occurring spontaneously or produced by activation protocols suggests the presence of heteroploidy in a high number of embryos.

Recently, it has been found that application of ultrasound enables intracellular uptake of drugs, macromolecules, DNA and fluorescent markers that are otherwise not permeable through the intact cell membrane. Ultrasound exposure is thought to induce the formation of pores in the cell membrane that permit easier inward transport of molecules or other agents in to the cells across the membrane barrier, resulting in faster and direct intracellular uptake. The work on these lines has been initiated in this institute under a NAIP funded research project.

Applications of Parthenogenesis

- Parthenogenetic embryos with their two sets of equivalently imprinted chromosomes are very useful models for the investigation of genome imprinting and its role in ontogenesis.
- Mammalian parthenotes serve as ideal models for studying the function of the maternal centrosome.
- The development potential of parthenogenetic embryos up to blastocyst is higher than SCNT produced embryos.
- Multiplication of superior animals without intervention of male germplasm.
- Cryopreservation of parthenogenetic embryos for conservation of germplasm of endangered animal.

Livestock Genomics and Marker Assisted Selection (Use of DNA Level Markers)

The advent of methods for visualization of polymorphisms at DNA sequence level and the development of powerful experimental designs have increased the practical potential for marker based approaches to improvement of production and health related traits in livestock. Development of the requisite molecular probes is required before marker based experiments can realistically be utilized for genetic improvement of livestock. With the demonstration of widespread molecular genetic polymorphism within populations and the association of certain markers with production related traits, experimental protocols are being developed for detecting the segregation of Quantitative Trait Loci , with the objective that genetic progress could be achieved through marker-assisted selction (Hallerman, 1989). Biotechnological methods allow the dissection of production traits into their individual Mendelian genes. For polygenic traits, to which the majority of production traits belong, these genes are referred as QTL. Identification of QTL will be useful in marker assisted selection schemes. Mapping genes explaining breed differences for economically important traits will all their introgression in other populations by marker aided backcrossing, thereby increasing the genetic variation usable as substrate for selection programmes. For example, high prolificacy of Chinese pig breeds and Boorola sheep breed can be used for improving the prolificacy of other breeds.

Genetic markers allow quantification of genetic diversity existing between domestic livestock populations which are essential for the implementation of cost effective conservation strategies.

One of the areas which can assist in genetic improvement of livestock is the study of **Major Histocompatibility system** in domestic animals. In experimental animals, an association has been demonstrated between immune responsiveness and susceptibility or resistance to diseases. Products of the major histocompatibility complex (MHC) mediate the immune response. In humans, the MHS is knowns as HLA (human leucocyte antigen system). In mice, the MHS is known as H-2(Histocompatibility complex 2). The serological identifications of goat lymphocyte antigens were initially demonstrated by van Dam *et al.,* 1979; Ruff and Lazary, 1985 and Nesse and Larsen, 1987. At least two loci coded for class I antigen. The molecules encoded by class I and class II genes are essential for the recognition of foreign antigens by the antigen receptors of T lymphocytes. The action of these genes has been believed to be responsible for observed association between the MHS, disease prevalence and immune responsiveness. The MHC has also been linked to reproduction in several species suggesting additional utility for MHC markers (Hallerman, 1989).

The study of MHC can lead to identification of genetically susceptible or resistant animals which should prove valuable in the testing of vaccines and therapeutic agents and aid in the breeding of disease resistant stock,

The biotechnology of rDNA can also help in the alteration of MHC in domestic animals in order to produce disease free or disease resistant farm animals in future.

Stem Cells

Stem cells are cells that have the ability to self replicate and give rise to specialized cells. Stem cells can be found at different stages of fetal development and are present in a wide range of adult tissues. Many of the terms used to distinguish stem cells are based on their origins and the cell types of their progeny.

There are three basic types of stem cells. Totipotent stem cells, meaning their potential is total, have the capacity to give rise to every cell type of the body and to form an entire organism. Pluripotent stem cells, such as embryonic stem cells, are capable of generating virtually all cell types of the body but are unable to form a functioning organism. Multipotent stem cells can give rise only to a limited number of cell types. For example, adult stem cells, also called organ- or tissue specific stem cells, are multipotent stem cells found in specialized organs and tissues after birth. Their primary function is to replenish cells lost from normal turnover or disease in the specific organs and tissues in which they are found.

Totipotent stem cells occur at the earliest stage of embryonic development. The union of sperm and egg creates a single totipotent cell. This cell divides into identical cells in the first hours after fertilization. All these cells have the potential to develop into a fetus when they are placed into the uterus. The first differentiation of totipotent cells forms a hollow sphere of cells called the blastocyst, which has an outer layer of cells and an inner cell mass inside the sphere. The outer layer of cells will form the placenta and other supporting tissues during fetal development, whereas cells of the inner cell mass go on to form all three primary germ layers: ectoderm, mesoderm, and endoderm. The three germ layers are the embryonic source of all types of cells and tissues of the body. Embryonic stem cells are derived from the inner cell mass of the blastocyst. They retain the capacity to give rise to cells of all three germ layers. However, embryonic stem cells cannot form a complete organism because they are unable to generate the entire spectrum of cells and structures required for fetal development. Thus, embryonic stem cells are pluripotent, not totipotent, stem cells.

Embryonic germ (EG) cells differ from embryonic stem cells in the tissue sources from which they are derived, but appear to be similar to embryonic stem cells in their pluripotency. Human embryonic germ cell lines are established from the cultures of the primordial germ cells obtained from the gonadal ridge of late-stage embryos, a specific part that normally develops into the testes or the ovaries. Embryonic germ cells in culture, like cultured embryonic stem cells, form embryoid bodies, which are dense, multilayered cell aggregates consisting of partially differentiated cells. The embryoid body derived cells have high growth potential. The cell lines generated from cultures of the embryoid body cells can give rise to cells of all three embryonic germ layers, indicating that embryonic germ cells may represent another source of pluripotent stem cells.

Much of the knowledge about embryonic development and stem cells has been accumulated from basic research on mouse embryonic stem cells. Since 1998, however, research teams have succeeded in growing human embryonic stem cells in culture. Human embryonic stem cell lines have been established from the inner cell mass of human blastocysts that were produced through *in vitro* fertilization procedures. The techniques for growing human embryonic stem cells are similar to those used for growth of mouse embryonic stem cells. However, human embryonic stem cells must be grown on a mouse embryonic fibroblast feeder layer or in media conditioned by mouse embryonic fibroblasts. Human embryonic stem cell lines can be maintained in culture to generate indefinite numbers of identical stem cells for research. As with mouse embryonic stem cells, culture conditions have been designed to direct differentiation into specific cell types (for example, neural and hematopoietic cells).

Adult stem cells occur in mature tissues. Like all stem cells, adult stem cells can self-replicate. Their ability to self-renew can last throughout the lifetime of individual organisms. But unlike embryonic stem cells, it is usually difficult to expand adult stem cells in culture. Adult stem cells reside in specific organs and tissues, but account for a very small number of the cells in tissues. They are responsible for maintaining a stable state of the specialized tissues. To replace lost cells, stem cells typically generate intermediate cells called precursor or progenitor cells, which are no longer capable of self-renewal. However, they continue undergoing cell divisions, coupled with maturation, to yield fully specialized cells. Such stem cells have been identified in many types of adult tissues, including bone marrow, blood, skin, gastrointestinal tract and dental pulp, retina of the eye, skeletal muscle, liver, pancreas, and brain. Adult stem cells are usually designated according to their source and their potential. Adult stem cells are multipotent because their potential is normally limited according to their source and their potential. Adult stem cells are multipotent because their potential is normally limited to one or more lineages of specialized cells. However, a special multipotent stem cell that can be found in bone marrow, called the mesenchymal stem cell, can produce all cell types of bone, cartilage, fat, blood, and connective tissues.

Blood stem cells, or hematopoietic stem cells, are the most studied type of adult stem cells. The concept of hematopoietic stem cells is not new, since it has been long realized that mature blood cells are constantly lost and destroyed. Billions of new blood cells are produced each day to make up the loss. This process of blood cell generation called hematopoiesis occurs largely in the bone marrow. Another emerging source of blood stem cells is human umbilical cord blood. Similar to bone marrow, umbilical cord blood can be used as a source material of stem cells for transplant therapy. However, because of the limited number of stem cells in umbilical cord blood, most of the procedures are performed for young children of relatively low body weight.

Neural stem cells, the multipotent stem cells that generate nerve cells, are a new focus in stem cell research. Active cellular turnover does not occur in the adult nervous system as it does in renewing tissues such as blood or skin. Because of this observation, it had been a dogma that the adult brain and spinal cord were unable to regenerate new nerve cells. However, since the early 1990s, neural stem cells have been isolated from the adult brain as well as fetal brain tissues. Stem cells in the adult brain are found in the areas called the subventricular zone and the ventricle zone. Another location of brain stem cells occurs in the hippocampus, a special structure of the cerebral cortex related to memory function. Stem cells isolated from these areas are able to divide and to give rise to nerve cells (neurons) and neuron-supporting cell types in culture.

Stem cell plasticity refers to the phenomenon of adult stem cells from one tissue generating the specialized cells of another tissue. The long standing concept of adult organ specific stem cells is that they are restricted to producing the cell types of their specific tissues. However, a series of studies have challenged the concept of tissue restriction of adult stem cells. Although the stem cells appear able to cross their tissue specific boundaries, crossing occurs generally at a low frequency and mostly only under conditions of host organ damage. The finding of stem cell plasticity carries significant implications for potential cell therapy. For example, if differentiation can be redirected, stem cells of abundant source and easy access, such as blood stem cells in bone marrow or umbilical cord blood, could be used to substitute stem cells in tissues that are difficult to isolate, such as heart and nervous system tissue.

Pluripotent embryonic stem cells (ESCs) could potentially generate specific cell types for treating serious diseases. A major problem limiting the clinical use of ESCs is the potential for tissues derived from these cells to be rejected by receiving patients. The most attractive solution to this problem comprises transplanting tissues derived from ESCs genetically matched to each patient. Somatic cell nuclear transfer (SCNT), where an adult somatic cell is returned to a pluripotent state (a process called reprogramming) following transplantation to an enucleated oocyte, can be used to provide such cells, however, ethical and practical limitations associated with both oocyte donation and human SCNT raise serious concerns about the suitability of this method.The isolation and *in vitro* maintenance of bovine embryonic stem cell lines is important for overcoming some of the difficulties related to production of transgenic cattle, and may significantly improve cloning efficiency in the bovine.

The establishment of pluripotent embryo derived cell lines from the bovine, however, is more complex and less efficient than in mice.

Alternative approaches to reprogramming cells include 1) fusion of somatic cells with ESCs and 2) introduction of a few key pluripotent genes into the somatic cells, also know as induced pluripotent stem cells (iPSCs).

Vaccine for Fertility Control and Increase Prolificacy

Immunocontraceptive vaccination is a fast-moving area of vaccine research and development in the human and animal health areas, with a number of products for livestock and companion animals recently brought to the market.

Since before written history, humans have practiced neutering of animals used for food and transport and to be kept as pets. This has been termed "man's first attempt at bioengineering". Following the discovery of the reproductive hormone system, attempts were made to control reproduction by immunization against key hormones in both humans and animals. Many early attempts gave encouraging results but were variable due to a lack of knowledge of how to consistently elicit an effective neutralizing immune response against self-proteins.

There are two goals of reproductive control vaccines that can be simply categorized as (i) immunocontraception and (ii) immunoneutering. Immunocontraceptive vaccines aim to prevent either fertilization of the oocyte by sperm or implantation of the fertilized egg yet retain sexual behavior patterns and competition in mating; this approach is most suited to the control of feral animal pests and native wildlife. Immunoneutering vaccines aim to prevent all sexual behaviors in both male and female animals as well as controlling fertility; these outcomes are suitable for companion animals, livestock, and, in some instances, feral animal pest control.

Auto-immunization to Increase Prolificacy

There have been three fecundity vaccines for sheep that have been commercialized, all based on stimulating an immune response to the steroid androstenedione. Vaccination of ewes against this intermediate steroid leads to a reduction in estrogen levels, and estrogen (β-estradiol) has a negative-feedback effect on the production of follicle stimulating hormone. Thus, immunoneutralization of androstenedione leads to the increased production of follicle-stimulating hormone, and this has the effect of increasing the frequency of multiple ovulations. The immunogen in the first available vaccine, Fecundin, was polyandroalbumin that contained androstenedione linked to human serum albumin. Similar vaccines, Androvax and Ovastim, are now marketed in New Zealand and Australia, respectively. Vaccination is carried out 5 and 2 weeks before mating for 6 to 8 weeks; in subsequent years, a booster dose is given to the flock 2 weeks before mating. The claimed increase in twinning is about 20 to 25% across a flock of ewes. The actual increased yield of lambs achieved with Fecundin was variable, and the vaccine was withdrawn from market. For these vaccines that increase fecundity, the correct nutritional maintenance of ewes with twins is not always easily managed, and vaccination will not correct underlying fertility problems.

Biotechnology of Conservation of Livestock Breed Biodiversity

Livestock breeds are recognized-I as important components of world biodiversity, because the genes and-gene combinations they carry may be useful to agriculture in the future. The gains in economic efficiency, which may result from using this genetic material, could far outweigh the costs of conserving the breeds in question II. Many breeds, once economically important, are now very rare, and yet they possess characteristics of potential value. A breed is 'a group of animals selected by man to have a uniform appearance that distinguishes them from other members of the same species'; worldwide, 3213 breeds of ass, cattle, goat, horse, pig, sheep and water buffalo have been enumerated, and possibly 1000 breeds altogether are at risk of extinction. Threatened breeds can be protected by *ex-situ* (cryogenic) or *in-situ* methods, the former being preservation as semen or embryos, the latter, the protection of breeding stocks. differentiation between livestock breeds has long been known to have occurred at nuclear loci 6 and phenograms of cattle breeds that accord with known or hypothesized history have been constructed using allele frequencies of biochemical polymorphisms. It is likely that studies evaluating-biological distinctiveness using autosomal polymorphisms will be used as criteria for conservation decisions. Autosomal microsatellites are highly polymorphic and are the markers of choice for modern genome mapping studies. They are also sensitive tools for the detection of genetic variation and are becoming widely used in practical population genetic applications 28 as well as in studies of livestock. The study of indigenous breeds and description of their unique genetic resources in the language of molecular genetics will soon provide guidance for decision makers. Mitochondrial DNA (mtDNA) analysis, which has proved extremely effective in the unravelling of genetic diversity in a wide range of organisms 24, has now been applied to cattle populations worldwide, and strongly implies that distinct races of the wild ancestor of cattle, the aurochs (Bos primigenius), were domesticated in at least two separate locales.

Gene for Prolificity: Booroola Merino

The Garole sheep from India is a prolific microsheep found in the hot, humid rice paddies of the southern part of West Bengal, particularly in the coastal belt of Sundarbans. These sheep average only 10 to 14 kg adult live weight and have a reported mean litter size of 2.27 in the Sundarbans region but have a lower mean litter size of 1.74 in the semiarid environment of the Deccan plateau of Maharashtra. Turner (1982) suggested that the highly-prolific Booroola Merinos can be traced back to an early Australian flock known to include prolific Bengal sheep. In-1792, 10 Bengal ewes and 2 rams arrived into Australia from Calcutta, and a further shipment of about 100 Bengal sheep arrived in the following year. Through the Australian-Stud Merino Flock register, Turner (1982) established a possible early link between Booroola and Bengal sheep; studies of the inheritance patterns of ovulation rate and litter size in prolific flocks have shown that major genes for prolificacy are segregating in Booroola (FecB) and Inverdale (FecXI) sheep. DNA Tests in prolific sheep from eight countries provide new evidence on origin of the Booroola (FecB) mutation. Recent discoveries that high prolificacy in sheep carrying the Booroola gene (FecB) is the result of a mutation in the BMPIB receptor and high prolificacy in Inverdale sheep (FecXI) is the result of a mutation in the BMP15 oocyte derived growth factor gene have allowed direct marker tests to be developed for FecB and FecXI. These tests were carried out in seven strains of sheep (Javanese, Thoka, Woodlands, Olkuska, Lacaune, Belclare, and Cambridge) in which inheritance patterns have suggested the presence of major genes affecting prolificacy and in the prolific Garole sheep of India, which have been proposed as the ancestor of Australian Booroola Merinos. The FecB mutation was found in the Garole and Javanese sheep but not in Thoka, Woodlands, Olkuska, Lacaune, Belclare, and Cambridge sheep. None of the sheep tested had the FecXI mutation. These findings present strong evidence to support historical records that the Booroola gene was introduced into Australian flocks from Garole (Bengal) sheep in the late 18th century. It is unknown whether Javanese Thin-tailed sheep acquired the Booroola gene directly from Garole sheep from India or via Merinos from Australia. The DNA mutation test for FecB will enable breeding plans to be developed that allow the most effective use of this gene in Garole and Javanese Thin-tailed sheep and their crosses.

Role of Biotechnology in Enhancing Production

Advances in understanding the regulation of nutrient use in agricultural animals have led to the development of technologies referred to as metabolic modifiers. Metabolic modifiers are a group of compounds that modify animal metabolism in specific and directed ways. They have the overall effect of improving productive efficiency (weight gain or milk yield per feed unit), improving carcass composition (lean: fat ratio) in growing animals, increasing milk yield in lactating animals, and decreasing animal waste per production unit (NRC 1994). Two classes of compounds have received major focus: somatotropins (STs) and β-adrenergic agonists. Somatotropin is a protein produced by the pituitary-gland that differs slightly in structure among animal species. Thus, commercial application of STs-depends on the use of recombinant DNA technology to produce the ST protein specific for a species. βadrenergic-agonists represent a class of compounds called phenethanolamines, and individual compounds differ in their biological effect. The several that affect animal growth are often referred to as repartitioning agents. Metabolic modifiers are considered to be new animal drugs and are regulated as such by the Food and Drug Administration's (FDA)-Center for Veterinary Medicine. The most commonly discussed ST is bovine somatotropin (bST), which has been administered to dairy cows to achieve increased milk yield, unprecedented improvements in productive efficiency (milk/feed), and decreased animal waste (Bauman, 1999). Specific biological mechanisms for the effects that bST has on nutrient use efficiency and animal wellbeing have been identified (Bauman and Vernon, 1993; NRC, 1994). Commercial sales of bST began in 1994, and use gradually has increased so that approximately half of U.S. dairy herds (> 3 million cows) are receiving bST supplements (Bauman, 1999). This usage involves herds of all sizes representing a full range of-management systems from all U.S. geographic regions. Commercial experiences have demonstrated further that bST supplements result in marked improvements in productive efficiency while maintaining normal cow health and herd life (Bauman, 1999; Etherton and Bauman, 1998). Bovine somatotropin is being used commercially in 19 countries. Experimental studies of ST use in growing animals have focused on porcine somatotropin (pST). Administration of pST to growing pigs results in greater nutrient use for lean tissue and less for body fat. This shift in nutrient partitioning results in substantially improved feed use, and the shift in lean:fat ratio represents an unprecedented improvement in carcass quality. Biological mechanisms that account for the effects of pST have been delineated and involve coordinated changes in lipid, protein, and carbohydrate metabolism (Etherton and Bauman, 1998; NRC, 1994).

Milk

Milk is produced by most mammals for the nurture of their young ones. Domestic animals have been selectively bred for centuries in such a way that the females produce milk far in excess of the requirements of their young ones. The excess milk is harvested by man for its own use. The animals which yielded most successfully to such genetic manipulation is the cow and buffalo. Although milk is also harvested from goat, sheep, camel, yak etc., but the major backbone of milk industry is cattle and buffalo milk. The milk produced by the mammary glands is harvested either by hand or machine milking through the teats. The production and let down of milk by the body of animals is controlled by hormones produced from various endocrine glands. The hormones are chemical substances produced by the endocrine glands in very small quantities but exert profound influence on the target glands through receptors present in the cells of the target gland.The study of hormones in lactation had its origin mostly in the beginning of 20 th century (Wilson, 2000). The early methods of study of role of various hormones in lactation included ablation of various endocrine glands and subsequent effects on milk yield and composition. For example, hypophysectomy results in failure of lactation indicated that anterior pituitary hormones are involved in lactogenesis (Cowie, 1969). Similarly surgical removal of thyroid gland reduces milk secretion indicating the role of thyroid hormone in lactation. However, the availability of bioassay techniques for measurement of hormones fuelled further researches in this field. The work of many a pioneers like Cowie, Turner, Tucker, Meites , Linzell, Peaker etc added new information to the role of hormones in lactation. The availability of sensitive radio immunoassay techniques in the 70s inspired a number of researchers to study the role of hormones in lactation qualitatively. Many a Indian scientists played a pioneering role in the study of lactation which included among others like Drs. Panda, Sud, Verman, Razdan, Ludri etc. The papers published by these workers are now classic. The use of hormone antagonists helped in studying the requirements of hormones during lactation. Administration of bromocryptine to goats resulted in decreased milk production (Singh and Ludri, 1998) indicating that prolactin is required for lactogenesis in early lactation in goats. Yash Pal and Ludri (1997) found that suppression of prolactin by bromocryptin in buffaloes resulted in low yields after parturition indicating the role of prolactin in milk secretion .Use of endocrinological research also opened up the possibility of artificial induction of lactation in animals suffering from reproductive disorders.

The availability of purified hormones in large quantities through biotechnology like bST, GHRH etc., brought to fore the practical applications of endocrinological aspects of lactation research. The use of bST indicated the possibility of increasing the milk production by about 25% through use of hormones (reviewed by Jindal *et al.*, 1989). Advances in biochemical studies gave us insight into the mechanism of synthesis of major milk constituents at the molecular level. The mechanism of protein, lactose and fat biosynthesis in the mammary cells was elucidated to a greater degree of precision by use of enzyme kinetics, tracer and other biotechniques. The role of somatic cell count in milk has been studied in relation to hormones to bring quality control aspects of milk production in the realm of lactation endocrinology. Somatic cell counts were found to be negatively correlated with milk yield during different stages of lactation and parity.

The current areas of lactation endocrinology include the role of several new hormones like Insulin like Growth factors (IGF) and several other such factors which were hitherto unknown but have profound influence on lactation of animals. In cattle and buffaloes, milk yield reaches a peak few weeks after parturition and then gradually declines. Lactation involves the role of growth hormone, prolactin, ACTH, TSH, insulin and parathyroid hormone etc. Studies on hormone concentration in serum with milk yield has shown the role of certain hormones in crossbred cattle and buffaloes and the association was studied through the use of correlation coefficients (Jindal, 1989; Jindal and Ludri, 1990 a,b, c, 1993, 1994, 1995). The role of growth hormone, insulin and thyroid hormone was elucidated during lactation. Singh and Ludri (1998, 1999, and 2000) and Yash Pal and Ludri (1997) elucidated the role of prolactin during early lactation in goats. Growth hormone is galactopoietic which is evident from numerous studies where growth hormone administered to animals increased milk yield upto 25% (Jindal *et al.*, 1989). Bauman and his colleagues have tried to explain the process as a homeorhesis phenomenon ie the nutrients are partitioned more towards milk production in case of growth hormone injected animals.The use of biotechnology in improvement of animal production in coming years were speculated earlier (Jindal and Madan, 1987). The use of scanning electron microscopy has also been attempted to facilitate the study of role of hormones in the synthesis of various milk constituents (Dang and Ludri, 2001). The twenty first century will see the application of these technologies in the form of increased milk production from domestic animals. The availability of newer techniques like gene injection for growth and other hormones will see the use of genetically modified animals for not only increased milk production but in the form of molecular farming of proteins from milk. But these technologies have several safety and other aspects which need to be properly addressed before application of such techniques on large scale. The twenty first century is expected to see increased research activity and greater dividends from such research.

Biotechnology of Rumen Manipulation

The major feed resources for ruminants particularly in arid and semi-arid areas are fibrous residues from-cereal crops and pasture or cut grasses from-wastelands. Usually both sources of ruminant diets are low in nitrogen (N) and digestibility. The nutritive value of the feed consumed and the efficiency with which rumen microorganisms convert dietary carbohydrate to end product of fermentation i.e., Volatile-Fatty Acids (VFA), determines the nutrient supply to the ruminant animal. Therefore, maximizing efficiency of breakdown and digestion of plant cell walls in the rumen can have a marked effect on animal productivity.

Present methods for manipulating ruminal fermentation that involve microbial biotechnology include dietary ionophores, antibiotics, and microbial feed additives. Several methods are currently employed to manipulate rumen fermentation to enhance post ingestion nutritive value of fibrous forages through use of biotechnology including inoculants of native and recombinant rumen-microorganisms, natural adaptation and microbial feed enzymes.

Developments in recombinant DNA technology mean that future methods will have a much wider scope. It has been suggested that genetically engineered ruminal microorganisms will be used in future to improve ruminal fermentation. *Butyrivibrio fibrisolvens, Ruminococcus* albus, *R. flavefaciens* and *Fibrobacter succinogeneses* are regarded as the primary fibre degrading bacteria in the rumen. There is considerable variation in fibre digesting capability of different strains of rumen microorganisms. Therefore, increasing the number of highly fibrolytic *Ruminococci* could potentially improve the rates of cell wall digestion in the rumen.

Genetically Modified Ruminal Microorganisms

With regard to fibre degradation in the rumen, much effort has been expended in developing genetically modified bacteria that would have superior fibre degrading abilities. The construction of genetically modified bacteria has proceeded under the assumption that the rumen microbiota does not produce the correct mixture of enzymes to maximize plant cell degradation. It is well established that the principal fibrolytic bacteria of rumen are *Ruminococcus* and *Fibrobacter,* but it is thought that they do not produce exocellulases that are active against crystalline cellulose, so that adding this activity would make them more potent. Ruminal bacterial species such as *Butyrivibrio fibrisolvens* and *Prevotella ruminicola* are found widely in ruminant animals on varied diets and are found in significant numbers regardless of the ruminal environment.

These species, therefore, are logical choices to introduce new or enhanced genetic material-into the rumen. Several technical objectives must be achieved before that will be possible. First, method for inserting foreign or modified genes into ruminal microorganisms and ensuring their efficient expression must be developed. Broad host range plasmids and transposons have been used successfully to introduce new DNA into ruminal bacteria, as have shuttle vectors constructed as chimeras of plasmids from ruminal species and *Escherichia coli.* Although so far only antibiotic resistance markers have been transferred, the prospects for introducing other genes into selected ruminal bacteria are excellent. Second, the expression of the gene products(s) should be known to be nutritionally useful *in vivo.* A few examples of this type of benefit have been demonstrated, and many more proposed, including polysaccharidases for improving fiber digestion, methods for improving the amino acid composition of ruminal bacteria, and breakdown of plant toxins. Third, introducing and maintaining the new strain in the mixed ruminal population is difficult. Factors governing the survival of new strains *in vivo* are ill understood, and attempts to select in favor of added new organisms have so far been unsuccessful. Because of the last obstacle, it may be advantageous, at least in the short term, to use nonruminal organisms, such as *Saccharomyces cerevisiae,* rather than indigenous ruminal species as a vehicle for implementing the benefits of recombinant DNA technology to ruminal fermentation. Yeast is already in widespread use as a feed additive, so no enrichment is necessary; and its genetics are already well known. Alternatively, adding particular enzymes to the diet may achieve some of the objectives described above, with the advantage that the manipulation could be achieved without the release of a recombinant microorganism. Paul *et al* (2003) have shown that fungal isolates from wild ruminants are superior to their counterparts in the rumen of domestic animals. When the isolates of rumen fungus from nilgai (*Basiolophus tragacamelus*) were inoculated in the rumen of buffaloes, the digestibility of fibrous feeds was significantly improved (Paul *et al.,* 2003; Sahu *et al.* 2004) showed that bacterial isolates from chinkara (*Gazella gazelle*) also improved *in vitro* and *in vivo* digestibility of wheat straw based diets. The results indicate that as soon as the inoculated microbes are withdrawn from the ration, the improvement observed in the nutrient utilization is also stopped, indicating poor survival of these microbes in the rumen as the guest microbes are not able to compete with the robust indigenous microbiota of the rumen (Kamra, 2008).

Over 100 different genes encoding enzymes for fibre digestion have been identified and cloned from ruminal bacteria such as *Butyrivibrio fibrisolvens, Fibrobacter succinogenes, Prevotella ruminicola, Ruminococcus albus* and *Ruminococcus flavefaciens*. At least 30 genes from ruminal fungi have been isolated that encode cellulase, xylanases, mannanases and endoglunases. Almost 50% of the fibrolytic genes cloned have been sequenced (Bowman and Sowell, 2003). These genes-are of particular interest due to their powerful fibrolytic activity and ability to break down very resistant cell wall polymers. Two plasmid vectors developed for use in other Gram positive bacteria have been introduced into four different *Ruminococcus albus* strains by electroporation (Cocconcelli *et al.*, 1992) and an efficiency of 3×10^5 transformants/µg was achieved with one of the plasmids, *Ruminococcus* strains did not persist for longer than 3 weeks before reaching undetectable levels. The resolution of this issue is fundamental to understand the true contribution that individual strains make to fibre degradation and to the population of species and genera in the rumen and overall fibre digestion. Compared with bacteria, the role of the fungi in rumen physiology is less understood. Five genera of anaerobic chhytridomycetous fungi have been isolated from the rumen: *Anaeromyces, Caecomyces, Neocallomastix, Orpinomyces* and *Piromyces* (Ho and Barr, 1995) and are considered to be involved in fibre degradation (Hespell *et al.*, 1997). Cellulases and xylanases are the enzymes produced by these fungi and are the most active fibrolytic enzymes (Gilbert *et al.*, 1992; Trinci *et al.*, 1994). Remarkably high activity xylanases have been isolated from *Neocallimastix patriciarum* and from *Orpinomyces joyonii* (Gilbert *et al.*, 1992; Li *et al.*, 1996). Genetically modified ruminal microorganisms: With regard to fibre degradation in the rumen, much effort has been expended in developing genetically modified bacteria that would have superior fibre degrading abilities. The construction of genetically modified bacteria has proceeded under the assumption that the rumen microbiota does not produce the correct mixture of enzymes to maximize plant cell degradation. A low frequency of transfer of the broad host range plasmid pAM$1 was also achieved into *R. albus* by conjugation from *Bacillus thuringiensis* BT351 (Aminov *et al.*, 1994). Despite these encouraging developments there have been no reports of introduction of vectors into *R. flavefaciens* strains and no studies on gene inactivation or the expression of genes introduced into *Ruminococcus*. Similarly there are no reports of indigenous plasmids or of successful attempts at gene-transfer into this important cellulytic species (Flint and Scott, 2000).

Reintroduction of natural and genetically modified microbes to the rumen: The ecology of the introduced strains has been overlooked to a large degree and the success of this technology may ultimately depend on a better understanding of factors that determine. Application of Biotechnology to Improve Post-ingestion Forage Quality in the Rumen-72-establishment in a complex microbial ecosystem, fermenting microorganisms (Durand and Fonty, 2001) survival and population density. Rumen biotechnology has the potential to improve the nutritive value of ruminant feedstuffs that are fibrous, low in nitrogen and of limited nutritional value for other animal species. Rumen biotechnology is defined as the application of knowledge of forestomach fermentation and the use and management of both natural and recombinant rumen microorganisms to improve the efficiency of ruminant production. Knowledge of rumen digestion that is relevant to modifying both the nutritive value of feedstuffs and the rumen microbial ecosystem by biotechnology is reviewed. Examples of the use and potential of biotechnology to alter the amount and availability of carbohydrate and protein in plants as well as the rate and extent of fermentation and metabolism of these nutrients in the rumen are discussed. The potential applications of biotechnology to rumen microorganisms are manifold and have been reviewed extensively.

Difficulties that are limiting factors in the progress of rumen biotechnology are:

- Isolation and taxonomic identification of strains for inoculation and DNA recombination;
- Isolation and characterization of candidate enzymes;
- Level of production, localisation and efficiency of secretion of the recombinant enzyme;
- Stability of the introduced gene;
- Fitness, survival and functional contribution of introduced new strains.

Ruminant animals develop a diverse and sophisticated microbial ecosystem for digesting fibrous feedstuffs. Plant cell walls are complex and their structures are not fully understood, but it is generally believed that the chemical properties of some plant cell wall compounds and the cross linked three dimensional matrix of polysaccharides, lignin and phenolic compounds limit digestion of cell wall polysaccharides by ruminal microbes. Three adaptive strategies have been identified in the ruminal ecosystem for degrading plant cell walls: production of the full slate of enzymes required to cleave the numerous bonds within cell walls; attachment and colonization of feed particles; and synergetic interactions among ruminal species. Nonetheless, digestion of fibrous feeds remains incomplete, and numerous research attempts have been made to increase this extent of digestion. Exogenous fibrolytic enzymes (EFE) have been used successfully in monogastric animal production for some time. The possibility of adapting EFE as feed additives for ruminants is under intensive study. To date, animal responses to EFE supplements have varied greatly due to differences in enzyme source, application method, and types of diets and livestock. Currently available information suggests delivery of EFE by applying them to feed offers the best chance to increase ruminal digestion. The general tendency of EFE to increase rate, but not extent, of fibre digestion indicates that the products currently on the market for ruminants may not be introducing novel enzyme activities into the rumen. Recent research suggests that cleavage of esterified linkages (e.g., acetylesterase, ferulic acid esterase) within the plant cell wall matrix may be the key to increasing the extent of cell wall digestion in the rumen. Thus, a crucial ingredient in an effective enzyme additive for ruminants may be an as yet undetermined esterase that may not be included, quantified or listed in the majority of available enzyme preparations. Identifying these pivotal enzyme(s) and using biotechnology to enhance their production is necessary for long term improvements in feed digestion using EFE. Pretreating fibrous feeds with alkali in addition to EFE also shows promise for improving the efficacy of enzyme supplements.

The potential use of recombinant DNA technology to modify fermentation characteristics of rumen microorganisms with the objective of improving efficiency of meat, milk or wool production by ruminant livestock has a great future. Some of the genetic modifications postulated are enhanced cellulolysis, reduced methanogenesis and decreased proteolysis/deamination. The technical difficulties involved in carrying out the gene modifications in terms of vector construction, uptake of DNA and expression of heterologous genes are yet to be taken care of . Finally, the feasibility of introducing a genetically altered microorganism into a complex symbiotic ecosystem is being questioned. Adding exogenous fibrolytic enzymes to ruminants diet can potentially improve cell wall digestion and the efficiency of feed utilization. Ultimately, the success of rumen biotechnology will depend on the environmental and regulatory concerns of the public being addressed.

Biotechnology of Animal Feeding

The largest impact of biotechnology on livestock production is increasing the livestock feeds value through improving nutrient content as well as digestibility of low quality feeds through use of efficient feed additives. One of the important area of biotechnology of animal feed and feeding is development of genetically modified feed ingredients and to improve certain feed ingredients, which have inherently low nutritional capabilities like high fiber, anti-nutritional factors, low proteins and deficiency of certain amino acids. Poor productivity of animals on straw based diets is a known fact (Kundu *et al.,* 1988). Chemical and physical attributes of straws that limit the digestion of cellulose and hemicelluloses include lignifications, silicification and crystallinity of cellulose. Recent advances in the understanding of lignin composition, polymerization, and regulation have revealed new opportunities for the rational manipulation of lignin in future bioenergy crops (Weng *et al.,* 2008). Some of the limitations which the nutritionist face during feed formulation are the antinutritive factors like trypsin inhibitors, saponins, tannins, phytates, oxalates, high fiber, limitations of phosphorus content in feed. Developing genetically modified feed having improved nutritional values could solve these problems. Microbial phytase as the result of biotechnology is an enzyme that breaks down the indigestible phytic acid (phytate) in cereals and oilseeds and release digestible phosphorus.

Probiotics

Probiotics produced from biotechnology can help to build up the beneficial bacteria in the intestine and competitively exclude the pathogenic bacteria. These bacteria also release enzymes, which help in the digestion of feed. The common organisms in probiotic products are

- *Aspergillus oryzae*
- *Lactobacillus bulgaricus*
- *Lactobacillus lactis*
- *Lactobacillus acidophilus*
- *Lactobacillus plantarum*
- *Lactobacillus salivarius*
- *Lactobacillus farciminis*
- *Bifidobacterium bifidium*

- *Streptococcus lactis*
- *Streptococcus thermophilus*
- *Saccharomyces cerevisiae*
- *Enterococcus faecium*
- In addition *E. faecalis, Bifidobacterium species, Bacillus subtilis, B. cereus, B. cereus toyoi, B. licheniformis, Pediococcus acidilactici* are also useful as probiotics.

The effects generally observed with probiotics in animal nutrition are increased productive parameters and better sanitary conditions. Lactic acid bacteria have the capability to enhance the immunity of the animals. Phagocytic activity and immunoglobulin levels have been increased by probiotics. *L. casei* and *L. plantarum* administered parenterally stimulate phagocytic activity. Yeast culture supplementation stimulates the growth of cellulolytic bacteria in the rumen. In ruminants, higher growth is positively related with the higher rumen fermentation pattern and nutrient utilization pattern. Probiotics, either lactic acid bacteria or yeast culture, have beneficial effect on higher nutrient utilization. Biotechnology based toxin binders will replace the present day binders like organic acids and their salts like propionic acid or adsorbents like bentonite, zeolites, hydroxyl aluminosilicates.

Biotechnology Crops as Animal Feed

Animals fed grains or forage produced via biotechnology do not differ in performance, yield or composition when the product is equivalent to its conventional counterpart for safety and nutrition. The same is true for animal products such as milk, meat and eggs. There have been questions that crops, which offer insect resistance and herbicide tolerance, affect the health, growth and performance of livestock? It has been concluded that crops developed via biotechnology are the same as conventional crops in nutrition, composition, safety and functionality in food and feed products. Transgenic crops have been fed to numerous species of livestock for several years. Studies comparing livestock fed transgenic and non-transgenic crops show that there is no difference between these types of crops in animal performance or composition of products from these animals. The results of these studies assure livestock producers that they can feed transgenic crops with confidence, knowing that the safety and nutritional value is comparable to the non-transgenic crops that have been fed to livestock for decades.

Biotechnology of Animal Products

As consumers increase demand for healthy food options, the meat animal industry will focus more strongly on improved meat composition. Biotechnology can lead to new and improved animal products. For example, it can modify the composition of milk, or the fat content of meat. Genetically transformed cows can produce designer milks with superior properties for use in various milk products. Added caseins in milk, for instance, can enhance cheese making. Increasing the phosphate group in casein can enhance the level of calcium. Removal of the source of lactose intolerance in milk can have a significant impact on the market for dairy products, especially among the 90 % of people with an Asian or African background who are lactose intolerant.

Progress in animal nutrition, reproduction, quantitative genetics, and the development of molecular genetics, proteomics, and functional genomics open new perspectives for the meat sector. The most promising developments include a wider utilisation of molecular markers, the possibilities of semen sexing and the targeted use of nutrition to modify the composition of meat. The increased use of biotechnology will have a considerable impact on the economics of production of meat and further processed products. New technologies will increase the possibilities for product differentiation and improve homogeneity of live animals.

Biotechnology can help address some of the meat safety issues and related challenges include the need to control traditional as well as "new," "emerging," or "evolving" pathogenic microorganisms, which may be of increased virulence and low infectious doses, or of resistance to antibiotics or food related stresses. Other microbial pathogen related concerns include cross contamination of other foods and water with enteric pathogens of animal origin, meat animal manure treatment and disposal issues, foodborne illness surveillance and food attribution activities, and potential use of food safety programs at the farm. Other issues and challenges include food additives and chemical residues, animal identification and traceability issues, the safety and quality of organic and natural products, the need for and development of improved and rapid testing and pathogen detection methodologies for laboratory and field use, regulatory and inspection harmonization issues at the national and international level, determination of responsibilities for zoonotic diseases between animal health and regulatory public health agencies, establishment of risk assessment based food safety objectives, and complete and routine implementation of HACCP at the production and processing level on the basis of food handler training and consumer education. Viral pathogens will continue to be of concern at food service, bacterial pathogens such as *Escherichia coli*, *Salmonella* and *Campylobacter* will continue affecting the safety of raw meat and poultry, while *Listeria monocytogenes* will be of concern in ready to eat processed products. These challenges become more important due to changes in animal production, product processing and distribution; increased international trade, changing consumer needs and increased preference for minimally processed products, increased worldwide meat consumption, higher numbers of consumers at risk for infection, and, increased interest, awareness and scrutiny by consumers, news media, and consumer activist groups.

Some New Products made Possible by Biotechnology Include

- The use of milk clottingenzyme and use of lipase in dairy industry is widespread. Biotechnologically produced microbial rennet can be used in place of animal rennet to coagulate milk during the making of cheese. Development of fermentation produced chymosin, which is produced by genetically modified organism with a gene encoding chymosin, is one example of milk clotting enzyme produced using genetic engineering (Reviewed by Kilcawley, 2006)
- New lactic acid bacterium produced by cell fusion technique;
- Safe anti-microbial agent, made from skim milk and glucose, which prolongs the shelf life of fresh milk;
- Biologically active peptides extracted from animal blood at slaughterhouses. These can be used as an additive in functional foods to improve human health;
- Colorant extracted from animal blood, to partially replace nitrite in meat products.

The quality and value of the carcass in domestic meat animals are reflected in its protein and fat content. Fat deposition and composition phenotypes are typically quantitative traits in nature, which are usually controlled by multiple genes and environmental factors. During the past 15 yr, both genome scan and candidate gene approaches have been widely used to localize chromosome regions or functional genes that affect these economically important traits. It has been widely believed that the application of these genes/markers in livestock breeding programs will permit the development of new genetic technologies and open the way to realize the full genetic potential for improvement of meat production for maximum profits.

To date, more than 300 genes and 400 QTL have been placed on the human obesity gene map . The map provides us a unique reference to explore genes/QTL for fat deposition and composition in livestock species.

People who suffer from food allergies manage their condition by avoidance strategies such as diet eliminations and careful examination of ingredient labels. Unexpected exposures and resulting adverse reactions to food allergens represent the main challenge in food atopy. Unlike hay fever and respiratory allergies, immunotherapy has achieved only limited success because of the potency of food allergens i.e., immunotherapy with food allergens can often trigger serious side effects, including anaphylaxis. Biotechnology had a positive influence on the science of food allergy by facilitating the discovery and characterization of allergens using recombinant DNA methods. Today, it is generally accepted that most major allergens have been identified and described. Subsequently, biotechnology has enabled the development of diagnostics based on recombinant allergens and more recently has been used to engineer potentially safer immunotherapeutic versions of food allergens, the creation of de-allergenized variants. This will allow safer immunotherapies since the de-allergenized variants of food allergens should produce fewer, if any, side effects. In addition, DNA vaccines based on this variants are being tested presently, with a view of providing long lasting immunotherapeutic options for food allergy patients. Biotechnology is also providing prophylactic options through the development of hypoallergenic-foods which have either been engineered to contain fewer endogenous allergens, or have been modified by the presence of proteins like theoredoxin, to render endogenous food allergens less potent and less allergenic. Hypoallergenic foods could reduce the incidence of new food allergies on a global basis.

3

Role of Biotechnology in Animal Health

Disease is an economically important issue in livestock production and has impact on human health, livelihood security of farmers and food safety. Infectious diseases caused by mycotic, parasitic, bacterial or viral agents are contantly threatening both the human and animal health. Biotechnology has been recognized as one of the tool in development of reliable and effective technique of diagnostic and vaccinology. Recent advancements in molecular biology have improved the existing technology to extent that can be used to produce purified immunogenic proteins essential for protective immunity.

Biotechnology has potential applications in the management of several animal diseases such as foot and mouth disease, classical swine fever, avian flu and bovine spongiform encephalopathy. The most important biotechnology based products consist of vaccines, particularly genetically engineered or DNA vaccines. Gene therapy for diseases of pet animals is a fast developing area because many of the technologies used in humans clinical trials were developed in animals and many of the diseases of cats and dogs are similar to those in humans. RNA interference technology is now being applied for research in veterinary medicine.

"An ounce of prevention is worth a pound of cure".

Vaccines

The major goals of veterinary vaccines are to improve the health and welfare of companion animals, increase production of livestock in a cost effective manner, and prevent animal-to-human transmission from both domestic animals and wildlife. These diverse aims have led to different approaches to the development of veterinary vaccines from crude but effective, whole pathogen preparations to molecularly defined subunit vaccines, genetically engineered organisms or chimeras, vectored antigen formulations, and naked DNA injections. Vaccination is the use of vaccines to prevent specific diseases. Vaccines are used to stimulate an animal's immune system to produce the antibodies needed to prevent infection. Over 100 years ago, in 1885, Louis Pasteur developed a crude nerve tissue vaccine for the post-exposure treatment of rabies. This form of vaccination used dessicated infected tissue and was found to prevent rabies infection in a 9 year old boy named Joseph Meister. Following Pasteur's initial vaccine, inactivated vaccines were developed. This inactivated form of vaccine was produced by serial dilutions followed by sterilization with chemical agents that inactivated the virus.

Louis Pasteur

Vaccination involves active immunization against a variety of microorganisms or their components, with the ultimate goal of protecting the host against subsequent challenge by the naturally occurring infectious agent. The terms vaccine and vaccination were originally used only in connection with Edward Jenner's method for preventing smallpox, introduced in 1796. In 1881 Louis Pasteur proposed that these terms should be used to describe any prophylactic immunization. Vaccination now refers to active immunization against a variety of bacteria, viruses, and parasites. Vaccines contain antigens (weakened or dead viruses, bacteria, and fungi that cause disease and infection). When introduced into the body, the antigens stimulate the immune system response by instructing B cells to produce antibodies, with assistance from T-cells.

Successful veterinary vaccines have been produced against viral, bacterial, protozoal, and multicellular pathogens, which in many ways have led the field in the application and adaptation of novel technologies. These veterinary vaccines have had, and continue to have, a major impact not only on animal health and production but also on human health through increasing safe food supplies and preventing animal to human transmission of infectious diseases.

Veterinary v/s Human Vaccines

The process of developing veterinary vaccines has both advantages and disadvantages over human vaccine development. On the one hand, the potential returns for animal vaccine producers are much less than those for human vaccines, with lower sales prices and smaller market sizes, resulting in a much lower investment in research and development in the animal vaccine area than in the human vaccine area, although the complexity and range of hosts and pathogens are greater. For example, the market size for the recently launched human vaccine (Gardasil) against papillomavirus and cervical cancer is estimated to be greater than 1 billion U.S. dollars, while the most successful animal health vaccines (e.g., against foot-and-mouth disease [FMD] virus in cattle and Mycoplasma hyopneumoniae in pigs) enjoy a combined market size that is 10 to 20% of this figure. On the other hand, veterinary vaccine development generally has less stringent regulatory and preclinical trial requirements, which can make up the largest cost in human vaccine development, and a shorter time to market launch and return on investment in research and development. In contrast to human vaccine development, veterinary scientists are also able to immediately perform research in the relevant target species. This is an obvious advantage over human vaccine development, as experimental infections, dose-response studies, and challenge inoculations need not be carried out in less relevant rodent models (Meeusen *et al.*, 2007).

While veterinary vaccines comprise only approximately 23% of the global market for animal health products, the sector has grown consistently due mainly to new technological advances in vaccine development, the continuous development of drug resistance by pathogens, and the emergence of new diseases. Apart from improving animal health and productivity, veterinary vaccines have a significant impact on public health through reductions in the use of veterinary pharmaceuticals and hormones and their residues in the human food chain. Most live veterinary viral vaccines induce mild infections with live organisms derived from nontarget hosts or attenuated through passage in different cell line cultures or chicken embryos (eggs). The virtual eradication of rinderpest virus from the globe is widely believed to have been critically dependent on the use of the "Plowright" vaccine.

Modern Veterinary Vaccines

Identification of the protective viral antigens potentially allows their isolation and/or recombinant production so that they can be administered as safe, nonreplicating vaccines. However, as isolated antigens generally induce poor protective immunity, subunit vaccines usually require repeated administration with strong adjuvant, making them less competitive.

Dow Agro Sciences successfully registered the first plant-based vaccine for Newcastle disease virus in poultry in the United States in 2005. Recombinant viral HN protein was generated in plant cell lines via Agrobacterium transformation and could successfully protect chickens from viral challenge.

The availability of complete DNA sequences and a better understanding of gene function have allowed specific modifications or deletions to be introduced into the viral genome, with the aim of producing well defined and stably attenuated live or inactivated viral vaccines. Recombinant DNA technology has provided the means to produce large quantities of inexpensive vaccines, while a better understanding of the immune system has helped produce vaccines that do a better job of boosting the body's immune system. These engineered products are safer than traditional vaccines. Whereas conventional vaccines sometimes revert to virulent (disease causing) forms, the new vaccines can be engineered to eliminate this threat.

Trovac AI H5 is a recombinant fowlpox virus expressing the H5 antigen of avian influenza virus. Several vaccines are available based on inactivated adjuvanted formulations for equine herpesvirus type 1 and equine herpesvirus type 4, equine herpesviruses that are major causes of abortion and respiratory disease.

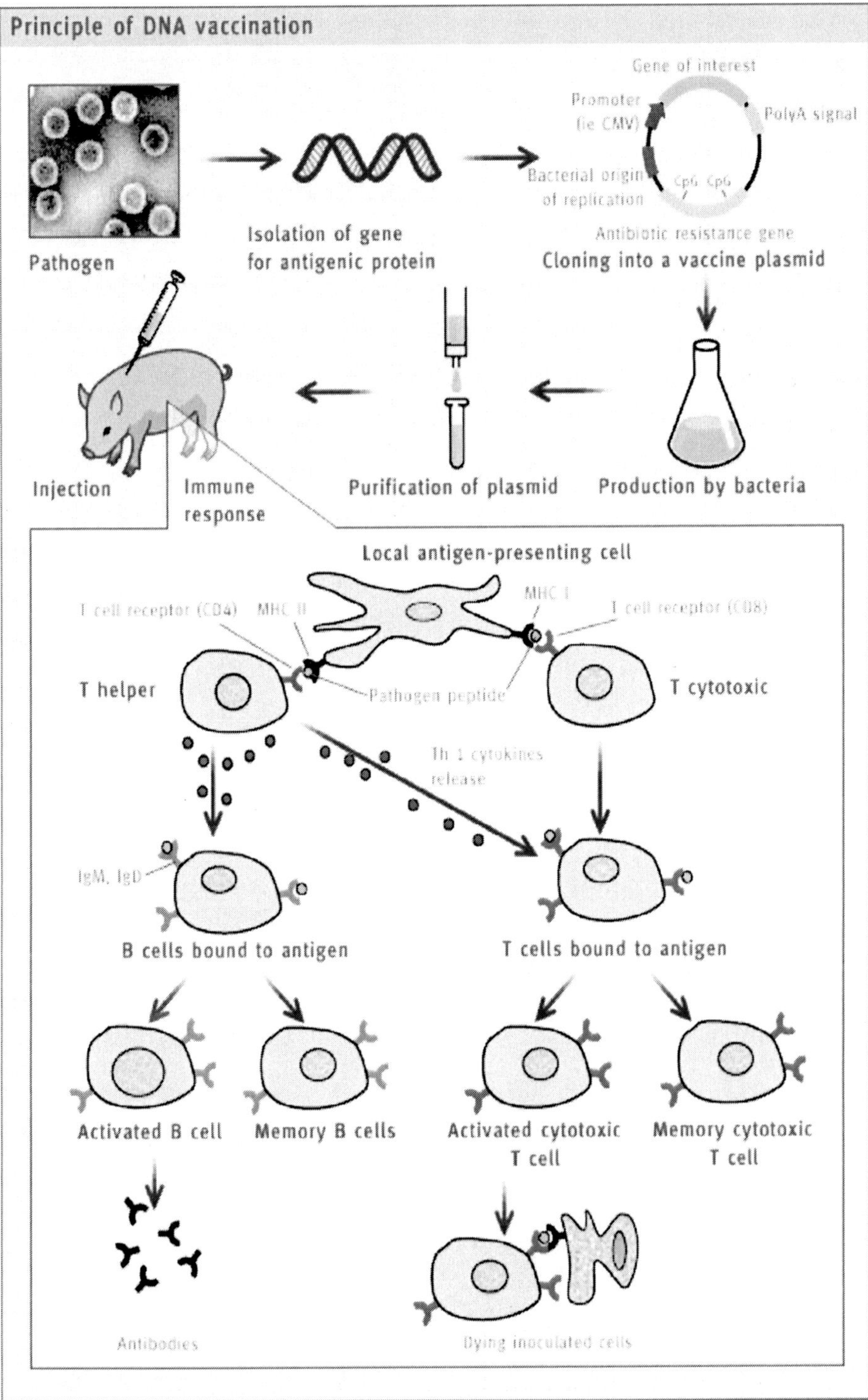
Principle of DNA vaccination
Gene of interest
Promoter (ie CMV)
PolyA signal
Bacterial origin of replication
CpG CpG
Antibiotic resistance gene
Pathogen
Isolation of gene for antigenic protein
Cloning into a vaccine plasmid
Injection
Immune response
Purification of plasmid
Production by bacteria
Local antigen-presenting cell
T cell receptor (CD4)
MHC II
MHC I
T cell receptor (CD8)
T helper
Pathogen peptide
T cytotoxic
Th 1 cytokines release
IgM, IgD
B cells bound to antigen
T cells bound to antigen
Activated B cell
Memory B cells
Activated cytotoxic T cell
Memory cytotoxic T cell
Antibodies
Dying inoculated cells

Immunization of animals with naked DNA encoding protective viral antigens would in many ways be an ideal procedure for viral vaccines, as it not only overcomes the safety concerns of live vaccines and vector immunity but also promotes the induction of cytotoxic T cells after intracellular expression of the antigens. Furthermore, DNA vaccines are very stable and do not require a cold chain. DNA vaccination of large animals has not been as effective as initially demonstrated in mice.

Many attenuated live or inactivated (killed) bacterial vaccines have been available for decades as prophylaxis against bacterial diseases in veterinary medicine. For most of the attenuated bacterial strains, the nature of the attenuation is not known, and since they have a proven track record, little is done to characterize underlying genetics. In some cases, however, the old and well-recognized live strains are not highly protective, and continued research is being performed to improve and develop new vaccines or vaccination strategies against, e.g., bovine tuberculosis, paratuberculosis, and brucellosis.

Gene deleted vaccines have been produced against strangles, a highly contagious disease in horses caused by infection with *Streptococcus equi* subsp. *equi*.

Brucellosis continues to be a major zoonotic threat to humans and a common cause of animal disease, especially in developing countries. Numerous attempts to produce a protective killed vaccine have so far been disappointing, and the most successful vaccines against brucellosis have been those employing live, attenuated *Brucella* spp.

Pigs are susceptible to both avian and human influenza viruses, and it is speculated that coinfection of pigs with highly pathogenic avian influenza virus and human influenza virus may create viral reassortant strains with the ability for human to human transmission.

rDNA Technology to Prepare Vaccines

The genetic engineering technique can also be used to prepare vaccines by deletion of gene of a virus that weakens the virus for producing the disease but preserve its immunogenicity. It was found that in an older attenuated pseudo-rabies viral vaccine one entire gene, responsible for causing diseases, was missing that made it suitable for a vaccine. Scientists at Baylor College of Medicine, Houston, USA, further disarmed the virus by removing part of another gene that codes for an enzyme that is needed by the virus to reproduce. The vaccine (omnivac) has been found to be immunogenic but quite safe for pigr. Practical application of genetic engineering also includes development of nucleic acid probes that are powerful tools for the early and rapid diagnosis of disease by nucleic acid hybridization (Cavanagh, 1986). Thus, the recombinant vaccines contain either a protein or gene coding for a protein of a pathogen origin that is immuogenic. The first recombinant antigen vaccine approved for human use was hepatitis B vaccine.

The advantage of rDNA technique is the efficient and economical production of relatively pure form of protein. For example the human insulin gene was used in the production of r DNA insulin and is now available commercially. Thus the problem of allergic reactions to bovine and porcine insulin in diabetics is eliminated.

There are limitless possibilities for preparing r DNA vaccines for both human and veterinary use. Instead of using vaccines that contain killed or attenuated live agents, subunit vaccines can be prepared. They consist of isolating the exact antigenic part of the agent responsible for producing neutralizing antibody and using it for immunization. The advantages of these vaccines are: elimination of viral mutation, infection of herd mates and vaccine derived illnesses. Recombinant subunit viral vaccine for acquired immuno deficiency syndrome virus in man (HIV, AIDS) may be the only hope for preparing vaccine for this dreadful disease because monantigenic HIV proteins are immuno suppressive. Another method of making vaccines is chemical synthesis of antigenic determinants which are proteins. If the amino acid sequence of the native antigenic determinant is known, it can be made synthetically (Bittle *et al.*, 1982).

The almost unbelievable will come when the rDNA technique is used to develop animals that are resistant to specific diseases or that have changed growth pattern or other characteristics making the animal more useful to us. Gene transfer combined with embryology has produced some spectacular experimental results (Palmiter *et al.*, 1982; Bandopadhyay and Temin, 1984; Crittenden and Salter, 1986). For examples a mouse embryo injected with a rat gene for growth has produced a 2 x mouse and the human growth hormone injected into a pig embryo has produced a very big pig. It is possible to transfer a horse gene to a cow that would make the cattle immune to foot and mouth disease. Is it possible to trick a cancer cell into accepting a dominant lethal gene that would prevent its growth and cure cancer?

Thus 'can you imagine' game could go on, but the process is extremely complex and is far from being a reality.

1. Recombinant DNA vaccines have been developed for the manufacture of rinderpest vaccine while sub-unit vaccine has been developed for FMD.
2. By using various membrane technologies for concentrating 146s particle of different seretypes in the FMD vaccine, it has now become possible reduce the dose of the vaccine.
3. Modified freeze-drying technology using right type of chemical stabilisers, the thermo-stability of rinderpest vaccine can be improved.
4. Through the technologies the extent of disease producing genes can be removed or replaced with other genes.

Biotechnology will certainly be helpful in curtailing the hectic vaccination schedule and in developing an effective and safe vaccine, which can provide solid immunity against variety of diseases.

Thus biotechnology plays a vital role in developing new vaccines of high safety. Recombinant vaccine and nuclic acid probes are also going to be a vital tools for future application in the field of diagnostics and vaccinology.

Veterinary Viral Vaccines

As there are no broad-spectrum antiviral pharmaceuticals available, hygienic measures to limit exposure and vaccination are the only means to prevent or control viral infections. Viruses (especially RNA viruses) are highly variable, and many viral infections are due to viruses with multiple serotypes (e.g., FMD virus, bluetongue virus, and influenza viruses). As a consequence, many of the existing viral vaccines are often unable to cope with the prevailing strains in the field, and new ones have to be generated from field strains with new outbreaks.

Live Vaccines

As the live organism can still infect target cells, these vaccines can replicate and induce both cellular and humoral immunity and generally do not require an adjuvant to be effective. Live products also offer the advantage of ease of administration, potentially in drinking water, intranasally, intraocularly, etc. However, they can pose a risk of residual virulence and reversion to pathogenic wild types as well as provide a potential source of environmental contamination. Despite such drawbacks of live viral vaccines, they have played a major role in successful disease control and eradication.

Killed Vaccines

Whole inactivated or killed viral vaccines are generally more stable and do not pose the risk of reversion to virulence compared to live vaccines, but their inability to infect cells and activate cytotoxic T cells makes them much less protective. Consequently, they generally require strong adjuvants and several injections to induce the required level of immunity and are usually effective in controlling only clinical signs rather than infection. Viral inactivation is commonly achieved through heat or chemicals (e.g., formaldehyde, thiomersal, ethylene oxide, and β-propriolactone). Much of the recent research in this area has concentrated on the development of improved adjuvanted formulations to overcome the effects of maternal antibodies on young animals.

Genetically Engineered Viral Vaccines

An interesting development in genetically engineered viral vaccines is the production of chimera viruses that combine aspects of two infective viral genomes. Poxviruses including vaccinia virus, fowlpox virus, and canarypox virus have been used as vectors for exogenous genes, as first proposed in 1982, both for the delivery of vaccine antigens and for human gene therapy. Poxviruses can accommodate large amounts of foreign genes and can infect mammalian cells, resulting in the expression of large quantities of encoded protein. A further application of vectored vaccines is the use of an attenuated viral pathogen as the vector, with the aim of inducing protection against two diseases, as with the live recombinant vaccine against both Marek's disease virus (MDV) and infectious bursal disease virus (IBDV) in chickens (Vaxxitek HVT + IBD).

Wider application of DNA vaccines will require further improvements and optimization for each host pathogen combination.

Veterinary Bacterial Diseases Vaccines

Many attenuated live or inactivated (killed) bacterial vaccines have been available for decades as prophylaxis against bacterial diseases in veterinary medicine.

Traditionally, attenuation of bacteria for the preparation of live vaccines has been performed by multiple passages in various media in the hope that some random mutation would deliver a nonvirulent, but replicable, type of the agent. With currently used molecular methods, the obtained deletions/mutations can be identified, but this technology also allows a more targeted design of live vaccines with specific deletions of predetermined known genes.

Veterinary Parasitic Vaccines

Protozoal infections in animals cause significant production losses and are a major impediment to the introduction of high productivity breeds in poorer, mainly tropical areas around the world. The most important veterinary trematode species are liver flukes (*Fasciola hepatica* and *Fasciola gigantica*). The chemical control of gastro-intestinal worms of ruminants poses a great threat to animal welfare and production. The repeated dosing of whole herds with synthetic anthelmintics is not sustainable as they produce food with drug residue, environmental pollution and anthelmintic resistance (AR). Vaccines for veterinary helminthes have focused on identifying protein antigens, which could be formulated as protective vaccines. Notable successes have been achived for some cestode parasites where recombinant protein have been developed into highly effective vaccine (Hein and Harrison, 2005). Increasing evidence suggest that parasite glycan moieties may provide an alternative source of vaccine antigen. The role of molecular biology in the development of vaccines against nematodes has been reviewed by Knox *et al.* (2001) and Lightowlers *et al.* (2003).Vaccine development against these parasites is hindered by the fact that they do not seem to induce immunity in their natural ruminant hosts, even after repeated infections. Recently, a unique breed of sheep (Indonesian thin tail) was shown to develop immunity against *F. gigantica*, and further dissection of this protective mechanism may offer new approaches to vaccine development.

Biotechnology is also producing an entirely new use for vaccines. They are being used to modulate hormones to increase growth rates, improve the efficiency of feed conversion, stimulate milk production, contribute to improved carcass quality and leaner meat, and enhance or suppress reproductive functions.

A number of improved vaccines for poultry have also been developed to protect birds from Newcastle disease, fowl cholera and infectious coryza.

DNA vaccine with an encoded target gene is being used to produce new vaccines. Tests have shown that the DNA vaccines consistently induced antibody response, and were resistant to toxin challenge.

DNA vaccine also have other advantages.

- They offer protection against diseases for which no vaccine is currently available.
- Their production does not need dangerous infectious agents.
- With mass production, they will not cost much to produce.
- Since they are stable at room temperature, storage costs will also be low.

Vaccines against cancer

Local vaccination with bacillus Calmette-Guérin (BCG) has long been used as a therapeutic treatment for superficial cancer of the urinary tract in humans and has been shown to be effective in the treatment of equine sarcoids and, to a lesser extent, bovine ocular squamous cell carcinoma. The mode of action of this vaccination regimen is unknown but may involve the upregulation of tumor-specific antigens through local inflammation and activation of the innate immune system (see review by Meeusen *et al,* 2007).

Some of the vaccines which have been developed using biotechnological tools are as follows:

Swine Flu

Swine Influenza is a highly contagious acute respiratory disease of pigs caused by type A influenza virus. Influenza in pigs is characterised by high morbidity and low mortality. Only type A influenza viruses are of significance to pig health. The most common type A Influenza subtypes that occur in swine are H1N1, H1N2 and H3N2.

- Fowl pox
- Haemorrhagic septicaemine in cattle and buffaloes
- Anti tick vaccine in dairy animals.

Foot and Mouth Disease

Foot and Mouth Disease (FMD) is a highly infectious viral disease of cloven hoofed animals. FMD is not transmissible to humans but the disease has major economic consequences. In the past FMD outbreaks in non-endemic countries have been controlled through the mass culling of infected and in contact animals. Currently FMD vaccines available are inactivated vaccines for the active immunization of cattle, buffalo, pigs, sheep and goats. A factory producing FMD Vaccine is located at Wagholi, a village near Pune. Now the the development of marker vaccines and test kits allows for the use of vaccination as part of a FMD control strategy. There is a FMD vaccine factory in Cologne which is controlled by Intervet.

Adenovirus Vector Vaccine

Replication defective human adenovirus vectors, which lack ability to replicate on their own, have been produced (Grubman *et al.,* 2003). These vectors can only grow productively in specific cell cultures that provide the missing functions, i.e. 293 cells. However, they can infect cells of several animal species, including cattle and swine. Used P1-2A3CD region of FMD virus, these adenovirus (HAd5) have been tested successfully in cattle. The vaccine holds immense potential.

Synthetic DNA vaccine against FMD will be highly preferable over the conventional vaccine, since the production of these vaccines donot require the handling of large amounts of live virus.

Johne's Disease

Paratuberculosis, which is also known as Johne's disease, is a chronic, progressive enteric disease of ruminants caused by infection with *Mycobacterium paratuberculosis*. Johne's disease is an infectious fatal wasting disease of sheep and goats. The clinical disease is characterized by chronic or intermittent diarrhea, emaciation, and death. Animals with subclinical paratuberculosis may cause economic losses because of reduced milk production and poor reproductive performance. Initially a live attenuated vaccine was used to control John's disease, which consisted of two live attenuated strains of *Mycobacterium paratuberculosis* suspended in a mixture of liquid paraffin, olive oil and pumice powder (Saxegaard and Fodstad, 1985).

Gudair vaccine is an animal vaccine for the control of Johne's disease in sheep and goat, which a Oil-water emulsion with vaccine manufactured by Pfizer. Scientists at Central Institute for Research on Goats have developed a Indigenous JD vaccine using native 'Indian Bison Type' MAP strain which has shown both therapeutic and prophylactic effect (Singh *et al.* 2008).

Diseases of Poultry

Newcastle Disease in Poultry

Newcastle disease is a highly contagious viral disease of domestic poultry, cage and aviary birds and wild birds. It is characterised by digestive, respiratory and/or nervous signs. The disease has a number of strains that differ in the severity of their clinical signs, ranging from inapparent infection to a rapidly fatal condition. Newcastle disease virus can infect many species of domestic and wild birds. Most susceptible are domestic fowls, turkeys, pigeons and parrots. Milder disease is seen in ducks, geese, pheasants, quail, guinea fowl and canaries. Clinical signs in poultry range from a mild, almost inapparent respiratory disease to a very severe depression, drop in egg production, increased respiration, profuse diarrhoea followed by collapse, or long-term nervous signs (such as twisted necks), if the birds survive. Death rate can be up to 100 per cent in severe forms of the disease. Newcastle disease is caused by a paramyxovirus. Presently lentogenic and mesogenic virus vaccines of chick embryo origin are being used for immunizing poultry against this disease. The vaccines in use are

- Poulvac Newcastle iK Vaccine (Inactivated)
- Websters Newcastle Disease Vaccine "V4 Strain" SPF (Living)
- Nobilis Newcavac Inactivated Newcastle Disease Vaccine
- Vaxsafe ND Vaccine (Living)
- Nobilis Gumboro+ND Combined Inactivated Vaccine
- Nobilis EDS+ND Combined Inactivated Vaccine

These vaccines suffer from following limitations.

1. Non availability of specific pathogen free (SPF) embryos for large scale vaccine production.
2. Risk of transmitting endogenous virus through chick embryos.
3. Serious reaction observed in vaccinates because of the use of whole embryo for the production of vaccine.

Avian Influenza (Bird Flu)

Avian influenza (which has been responsible for 'bird flu' outbreaks in recent years) is caused by an orthomyxovirus. H5N1 strain of Avian Influenza (AI, or Bird Flu) has spread through Asia and to Europe. A recently developed vaccine against avian influenza virus

(Poulvac FluFend) is where the hemagglutinin (HA) gene has been removed from an H5N1 virus, inactivated by removing the polybasic amino acid sequences, and combined with the NA gene from an H2N3 virus onto an H1N1 "backbone" virus. A vaccine containing the resultant inactivated H5N3-expressing virus administered in a water-in-oil emulsion protects chickens and ducks against the highly pathogenic H5N1 strain.

Rabies

Today human diploid cell vaccines (HDCV), also an inactivated virus vaccine, are used almost exclusively by the developed world for pre and post vaccination of Rabies. Current human diploid cell vaccine (HDCV) in the market includes Rabipur (PCEC) from chicken embryos. DNA vaccines utilize naked DNA strands alone for immunization, without traditional proteins or carrier viruses. The genes encoding immugeneic proteins are inserted into a circle of bacterial DNA known as a plasmid. DNA vaccine uses just enough genes from the virus to activate the immune system. The plasmids are shot into muscle tissue using a gene gun, where they are taken up and expressed by cells. Surprisingly enough this method of immunization has been shown to work surprisingly well when it comes to eliciting an immune response against pathogens, yet it is still not clear exactly how the DNA vaccines stimulate an immune response. It must be also kept in mind that the response is still not as strong as that seen with traditional vaccines. Recently announced was the first DNA vaccine that was shown to prevent rabies in monkeys. All eight of the monkeys treated with the DNA vaccine before being given a lethal dose of the rabies virus survived. DNA vaccine is very inexpensive to make and it's very stable, so it does not require refrigeration. This makes it useful candidate in developing countries. But the new vaccine can only be used prior to exposure to rabies, and is not a replacement for post-exposure treatment with IgG and HDCV. Researchers said there are no immediate plans to conduct human clinical trials, due to questions concerning whether additional genes must be inserted to offer protection to humans, and whether it would be effective if it were administered after a person contracted rabies.

Disease Diagnostics

Diagnosis of infectious diseases is of paramount importance for the applications of preventive measures and therapy. The diagnosis technique has to be rapid, specific, sensitive and reliable. The infectious diseases are generally diagnosed by :

- Macroscopic or microscopic examinations of tissues and body fluids.
- Culture methods
- Immunological methods like identification of antigens or measurement of specific antibodies.
- Isolation, identification and visualization of virus.
- Serological methods combined with neutralization, complement fixation, gel precipitation, haemoagglutination-inhibition, RIA and ELISA.

These methods have number of limitations and molecular techniques have recently become available which overcome some of the shortcomings by being more specific and accurate.

Molecular diagnosis is assuming an important place in veterinary practice. Polymerase chain reaction and its modifications are considered to be important. Fluorescent *in-situ* hybridization and enzyme-linked immunosorbent assays are also widely used. Newer biochip-based technologies and biosensors are also finding their way in veterinary diagnostics. Molecular diagnostics enables the early detection of diseases and conditions, the selection of targeted, individualised treatment options and the monitoring of treatment efficacy. Gen-Probe Incorporated, Innogenetics NV and Roche Molecular Diagnostics are few of the major immunodiagnostics companies in the world.

Molecular biological methods have become increasingly applicable to the diagnosis of infectious diseases. One important benefit from biotechnology is the diagnosis of livestock diseases, and genetically transmitted conditions which damage health and productivity. Molecular diagnostics is the fastest-growing segment of the *in vitro* diagnostics industry but still in nascent stage in veterinary diagnosis largely because of the huge cost involved.

Worldwide there are about 600 new biotechnology-based diagnostics in the market with a value of about 20 billion US $. Many more are about to enter the market, the most prominent among there will be PCR based diagnostics. In India, the diagnostic sales are expected to be between 1 to 2 billion rupees (Padha, 1996). India relies on imports for many of the immunodiagnostic kits. Many of the locally developed diagnostics have failed, while the imported diagnostics are either unsuitable or expensive.

The Nucleic Acid-hybridization Technique

In this technique the nucleic acid to be analysed is denatured and filtered through a solid matrix such as nitrocellulose or nylon membranes. The single strands bind to the filter in such a way that prevents self annealing and allows the sequences of interest to interact with nucleic acids in a solution that is in contact with the filter. The nucleic acid used to detect the presence of a specific nucleic acid in the immobilized sample is radioactively labeled and is known as the 'proble'. If complementarity exists between the probe and the immobilized nucleic acid, hybridization will occur and this can be measured by a scintillation counting or autoradiography. This technique is known as 'Dot-blot' or 'spot-blot' hybridization. The practical application of this technique was used for detection of herpes simplex virus in infected cells and specimens (*see review by Moreno-Lopez, 1988*).

Diagnostics Based on PCR

PCR is a highly sensitive procedure for detecting infectious agents in host tissues and vectors, even when only a small number of host cells are infected. PCR can target and amplify a gene sequence that has become integrated into the DNA of infected host cells. It can also target and amplify unintegrated viral gene sequences. However, PCR method does not differentiate between viable and nonviable organisms or incomplete pieces of genomic DNA, and this may complicate interpretation of results and affect the applicability of PCR in this role. PCR may prove to be very useful in the diagnosis of chronic-persistent infections, such as those caused by retroviruses (bovine leukaemia virus, caprine arthritis/encephalitis virus, etc.). These diseases present serious problems in terms of diagnosis and prevention since infected animals are a constant potential source for transmission. When PCR is used for diagnosis, a great deal of care is required to avoid contamination of the samples because the exquisite sensitivity of the technique can easily lead to false-positive results.

Classical PCR methods for diagnosis of pathogens, both bacterial and viral, are now being complemented and in some cases replaced with real-time PCR assays. Real time PCR monitors the accumulation of PCR product during the amplification reaction, thus enabling identification of the cycles during which near-logarithmic PCR product generation occurs. In other words, the assay can be used to reliably quantify the DNA or RNA content in a given sample. In contrast to conventional PCR, real-time PCR requires less manipulation, is more rapid than conventional PCR techniques, has a closed-tube format therefore decreasing risk of cross contamination, is highly sensitive and specific, thus retaining qualitative efficiency, and provides quantitative information.

Diagnosis by Restriction Fragment Length Polymorphisms and Related DNA-based Approaches

Serological tests that are commonly used to identify micro-organisms may not distinguish between isolates of closely related pathogens, whether they be viruses, bacteria, fungi or parasites. A DNA-based procedure will offer-the better discrimination that is often required and an appropriate starting point may be analyses for restriction-fragment length polymorphisms (RFLP). The RFLP approach is based on the fact that the genomes of even closely related pathogens are defined by variation in sequence. For example, the linear order of adjacent nucleotides comprising the recognition sequence of a specific restriction enzyme in one genome may be absent in the genome of a closely related strain or isolate. Reverse transcription to DNA and then digesting the nucleic acid with one of a panel of restriction enzymes, the individual fragments within the digested DNA are then separated within a gel by electrophoresis and visualised by staining with ethidium bromide. Ideally each strain will reveal a unique pattern, or fingerprint. Many different restriction enzymes may be considered at the outset of a new piece of work, so that analyses of many molecular fingerprints from digestions with several individual restriction enzymes may be undertaken and combination of the best set of results will allow a comprehensive differentiation between strains or isolates.

Diagnosis by DNA Probes and DNA Microarray Technology

In conventional DNA probing the detection of a pathogen is limited by the number of probes used, whereas in microarray analysis one is limited only by the number of target DNAs on the array.

Immunoblotting

Immunoblotting combines the high resolution of gel electrophoresis with the specificity of immunochemical detection and offers a means of identifying immunodominant epitopes recognised by antibodies from infected animals or MAbs directed against the target agent.

Antigen-Capture Enzyme-Linked Immunosorbent Assay (ELISA) Proteomics

Main diseases where these tools are being used

Foot and mouth disease
Rinderpest
Brucellosis
Tuberculosis
Johne's disease
Leptospirosis
Bovine viral diarrhea
Infectious boving rhinotracheitis
Rabies
Mycotoxeciosis

Infectious bursal disease
Marek's disease
New Castle disease
Infectious bronchitis
Theileriosis
Mycoplasmosis
Chlamydiosis
Rota, parvo and corona
Virus infections
Equine infectious anaemia

Disease Diagnosis Kits Developed

Brucelosis

Brucella melitensis causes a worldwide zoonosis. It is one of the major causes of abortion in goats and sheep and the organism is secreted in the milk of infected animals. Scientists at CIRG, Makhdoom have developed a PCR-RFLP for Molecular Typing of Brucella from Goats (Gupta, V.K.).

Rabies

Dipstick dot ELISA and dot ELISA have been standardized and field diagnostic kits based on these techniques have been developed and released.

Canine Distemper and Canine Parvo Virus

A rapid Dot ELISA method has been standardized for the detection of Canine distemper and Canine Parvovirus. Dipstick ELISA has been standardized for the detection of antibodies against Canine Parvovirus (CPV) and Canine Distemper Virus (CDV).

Canine Hepatitis Virus

Dot ELISA and Dipstick Dot Elisa have been standardized and field kits have been developed for the diagnosis of Infectious Canine Hepatitis (Canine Adevovirus type 1). Latex agglutination test has been standardized for quick diagnosis of Infectious Canine Hepatitis.

The developed rapid tests have been compared with laboratory Haemoagglutination test using Human '0' erythrocytes.

Foot and Mouth Disease

Foot-and-mouth disease, FMD or hoof-and-mouth disease (Aphtae epizooticae) is a highly contagious and sometimes fatal viral disease of cloven-hoofed animals, including domestic animals such as cattle, water buffalo, sheep, goats and pigs, as well as antelope, bison and other wild bovids, and deer. It is caused by foot-and-mouth disease virus. Humans are very rarely affected. There are seven FMD serotypes: O, A, C, SAT-1, SAT-2, SAT-3, and Asia-1. These serotypes show some regionality, and the O serotype is most common. In India, Endemic - Serotypes O,A and Asia1 are prevalent. On June 18, 1981, the U.S. government announced the creation of a vaccine targeted against FMD; this was the world's first genetically engineered vaccine. Bommeli Diagnostics, Switzerland, subsidiary of Intervet International, has developed Chekit-FMD-3ABC, an ELISA testkit, in collaboration with the World Reference Laboratory in Pirbright (U.K.) and the Instituto Zooprofilattico in Brescia (Italy).

New Castle Disease

An MPA-NDV ELISA kit was designed as a qualitative test for monitoring the vaccination responses and detecting the protective levels of antibody in a single serum dilution. The kit does not need tile use of ELI SA reader. The kit was evaluated in the field on a total of 813 serum samples representing 9 flocks (flock size ranging from 20,000- 2, 00,000 birds). The kit was found to be 100 percent sensitive and specific when compared with standard H.I. test.

Egg Drop Syndrome Virus-76 (EDS-76)

An EDS-ELISA kit based on a qualitative penicillinase ELISA with a single serum dilution was developed. The test was evaluated on a total of 500 serum samples comprising 176 serum samples from unvaccinated and 324 from vaccinated commercial layer flocks. The kit was 98.3 percent" sensitive and 100 percent specific for detecting EDS-76 antibodies. Apart from this a latex agglutination test was developed for detection of EDS-76 virus n field samples and a Multi-kit employing dot-immuno-binding assay (DIA) for detection of antibodies to NOV, EDS-76 and ISO virus has been developed.

Proteomics for Disease Diagnostics

Proteomic technologies will play an important role in drug discovery, molecular diagnostics and practice of medicine in the post-genomic era - the first decade of the 21st century. Most commonly used technologies are 2D gel electrophoresis for protein separation and analysis of proteins by mass spectrometry. Microanalytical protein characterization with multidimentional liquid chromatography/mass spectrometry improves the throughput and reliability of peptide mapping. Matrix-Assisted Laser Desorption Mass Spectrometry (MALDI-MS) has become a widely used method for determination of biomolecules including peptides, proteins. Functional proteomics technologies include yeast two-hybrid system for studying protein-protein interactions. Establishing a proteomics platform in the industrial setting initially requires implementation of a series of robotic systems to allow a high-throughput approach for analysis and identification of differences observed on 2D electrophoresis gels. Protein chips are also proving to be useful. Proteomic technologies are now being integrated into the drug discovery process as complimentary to genomic approaches.

Nanotechnologies in Diagnosis and Vaccine Development

Nanotechnologies involve working at the atomic, molecular and supra-molecular levels in the length scale of 1–100 nm range, in order to understand, create and use materials, devices and systems with fundamentally new properties and functions because of their small structure. Nanoparticles have exhibited-tremendous potential for detecting disease markers, pre-cancerous cells and fragments of viruses. Also, metal-coatings and metal nanoparticles functionalised with different biomolecules have been found useful in detecting-specific proteins and antibodies. For example, researchers have synthesised a specially charged silicon nano-wire-connected with an antibody receptor that can detect the presence of cancer markers in the blood even if the-concentration of these antigens in blood is about a hundred-billionth of the protein content. These sensors are-much more accurate than currently available technologies.

Monoclonal Antibodies

Body's own defence system guards both man and animals against various infectious and non infectious diseases. Besides protecting, the immune system also helps in recovery from a variety of diseases. This system is based on the production of antibodies or immunoglobulins in response to various antigens present in disease organisms. When an animal is exposed to a foreign substance (antigen) , its lymphocytes recognize the antigenic sites (epitopes) on the antigen and antibodies that bind with these sites are produced. Each responding lymphocyte divides into a clone of cells and each member of the clone produces identical antibody molecules. Antigens generally have many epitopes and many clones respond to the antigen by producing different types of antibodies. Normal antibody response of an animal to an antigen, therefore, is polyclonal because many clones produce antibodies of many different types. As such, most of these antibodies present in the body are polyclonal.

Since the discovery of monoclonal antibodies in 1975, for which Kohler and Milstein received the 1984 Nobel Prize, a new field in Biotechnology has been created (Kohler and Milstein, 1985). The basis of MAb production is to fuse 2 cells, an antibody producing lymphocyte and a myeloma or tumor cell, to form a hybird. The lymphocyte that normaliy produce antibodies, have predictable short life span. The myeloma cells donot produce antibodies but reproduce rapidly and are referred to as immortal cells. Both of these cells are placed in a solution to dissolve the cell and to encourage them to fuse, thus producing the hybrid. The hybridomas are separated, cloned and tested by very sensitive assays such as enzyme linked immunosorbent assay or radio immunoassays to determine which one produces the antibody to hit the target. They combine the most desirable charateristics of both the parental cell types; antibody production and immortality. Those that are acceptable are growth either in mice cultures producing large amounts of antibodies over long period of time. Because they recognise only one antigenic epitope, MAbs are always more highly specific and uniform than polyclonal antibodies (Purchase, 1986).

Applications of Monoclonal antibodies

These antibodies have many important uses. These include-

- *Diagnosis of disease-* A number of MAb diagnostic ELISA kits are now commercially available for early and rapid diagnosis of many animal diseases like colibacillosis, feline leukemia etc.

- Quality assurance of recombinant vaccines.
- Identification of antigenic sites involved in neutralization.
- Potential therapentic applications (Immuno therapy).
- 'Magic bullet' to kill specific cancer cells or viral antigens.
- *Anti-idiotypic vaccines:* In this complex technique, a MAb produced against the antigenic binding cleft (idiotype) of another MAb actually becomes the vaccine. Since many animal viral vaccines are live, the virus in these vaccines can potentially mutate and cause disease eg FMD. Through the use of anti-idiotypic vaccine, losses due to live attenuated vaccination would be decreased.

Disease Diagnosis : Monoclonal antibodies were initially used for disease diagnosis by differential diagnosis. Because of their great specificity, MAbs have proved-ideal for differential diagnosis between vaccine strains and field strains of organisms.

Immunotherapy

Because of their unlimited availability, MAbs have the most intriguing potential of being used in passive immunotherapy. The first commercial application of MAb administration *in vivo* was to protect neonatal calves against diarrhoea caused by enterotoxigenic *E.coli.*

Problems

One of the major limitations of MAbs for use in immunotherapy is their murine origin. When applied in other species, they are recognised as foreign, which limits their effectiveness. The production of homologous MAbs is harmpered by lack of myeloma fusion partners (e.g. Bovine or j.j. porcine origin) and unstability of interspecies heterohybridomas.

Because of their unlimited availability, MAbs have the potential of being used in passive immunotherapy. The first commercial application of MAb administration *in vivo* was to protect neonatal calves against diarrhoea caused by enterotoxigenic *E.coli.*

Monoclonal antibodies being pathogen specific are not effective in mixed infections. Monoclonal antibodies are more effective in preventing infections than in controlling exisiting diseases.

Monoclonal antibodies need to be administered systematically as being proteins in nature gets destroyed in the digestive tract. The cost of production of monoclonal antibodies is very high.

In a novel approach, mouse x bovine hybrid myelomas (heteromyelomas) have been constructed in attempt to obtain a better fusion partner for the production of bovine MAbs (Boman *et al.*, 1990). With these xeno hybridmas bovine MAbs have been produced against Forrsman antigen, Rotavirus, PMSG and K99 antigen of *E.coli.*

Monoclonal antibodies have found applications in anti-cancer therapy. Although monoclonal antibodies are widely used in human medicine, their use in veterinary medicine is still limited. Using this technology, animals with enhanced natural disease resistance will be identified as well as individuals that are specifically resistant to the most important infectious diseases, e.g. foot and mouth disease, avian influenza.

Animal Genomics for Disease Resistance

Infectious disease adversely affects livestock production and animal welfare, and has impacts upon both human health and public perception of livestock production. Breeding of farm animals for disease resistance is one of the long term goal of modern animal biotechnology. Researchers are debating the scientific views on genomic selection. There occurs a large variation in disease resistance among animals. There is well-documented evidence for between-animal genetic variation in resistance to the disease and, in the case of some of the infectious diseases, resistance to infection. These heritable differences between animals lead to opportunities to breed animals for enhanced resistance to the disease. Evidence has accumulated that local breeds are more disease resistant than imported breeds to a variety of pathogens (Spooner, 1997). There are many economically important diseases for which there is no cost effective vaccine e.g. Trypanosomiasis, African swine fever and Theileria sergenti infection. With some of these there is no treatment either, but even where there is a therapy, drug resistance becomes a problem. Development of farm animals for disease resistance can be a viable alternative wherein we can cash in on the variation in disease resistance between different breeds or even animals. Part of the natural variation seen for resistance to diseases in ruminants is under genetic control. Genetic method of controlling the disease seems to be a viable alternative as it is sustainable and can prevent environmental pollution and provides safer food and can enhance productivity. Identification of some of the genes involved in regulating resistance to diseases will allow early selection of genetically superior animals and may increase the rate of selection for resistance to diseases. Genetic resistance to disease is the sustainable control measure par excellence. Because once established it does not require high management inputs to maintain it. Sometimes vaccine and disease resistance has been used simultaneously to control health problems as has been the case of poultry (Spooner, 1997).

Despite increased biosecurity and improved genetic resources infectious diseases and metabolic disorders continue to cause major losses in our food production animals. The availability of the full genome sequence in chickens and rapid advances in cattle and swine genomics have opened up major opportunities. Because no one gene, or no one form of a gene, can confer resistance to all pathogens, even marker based selection for susceptibility or "resistance" may have to be balanced to ensure that a herd or a population will have an adequate immune response to a variety of pathogens. Alternatively, susceptibility to a specific pathogen, whose presence is confirmed by DNA based means, could be selected against while maintaining marker and allelic heterozygosity for general immune responsiveness and other production traits.

There are now concerted efforts to find genetic markers associated with resistance to infection, potentially allowing selection for increased resistance in the absence of infection. To identify the genes responsible for genetic variation in host susceptibility to pathogens, genetic markers that account for a significant portion of genetic variance need to be identified and incorporated into a routine screening procedure. A growing number of quantitative trait loci (QTLs) associated with production has been identified as a result of genetic mapping efforts in livestock. As the human genome effort increases scientific knowledge of the function of genes involved in the complex relationship between host and pathogen, researchers will be able to identify new DNA sequences associated with resistance or susceptibility and to produce genetic markers useful for selection in most, if not all, livestock species. Functional genomics, combining gene mapping and gene expression studies, is now being applied to a number of diseases, an example being the identification of genes whose expression underlies genetic differences in resistance to nematode parasites (Bishop and Morris, 2007). Such data on genetic selection for disease resistance will result in food animals with improved health and decreased need for antibiotic usage, thus benefiting consumers worldwide.

Gene Knockout Technology

Animal biotechnology also can knock out or inactivate a specific gene. Knockout technology creates a possible source of replacement organs for humans. The process of transplanting cells, tissues, or organs from one species to another is referred to as "xenotransplantation." Currently, the pig is the major animal being considered as a xenotransplant donor to humans. Unfortunately, pig cells and human cells are not immunologically compatible. Pig cells express a carbohydrate epitope on their surface that is not normally found on human cells. Humans will generate antibodies to this epitope, which will result in acute rejection of the xenograft. Genetic engineering is used to knock out or inactivate the pig gene (alpha1, 3 galactosyl transferase) that attaches this carbohydrate epitope on pig cells. Other examples of knockout technology in animals include inactivation of the prion related peptide (PRP) gene that may generate animals resistant to diseases associated with prions (bovine spongiform encephalopathy [BSE], Creutzfeldt-Jakob Disease [CJD], scrapie, etc.).

Gene Pharming

Pharming is a new concept where therapeutic drugs are produced in farm animals. The most commonly used animals for gene pharming are cows, sheep, goats, mice, pigs, chickens, rabbits, swines etc and these animals produce the proteins for medicine in their milk, urine, blood or eggs, which are later on purified. Such animals carrying human genes for therapeutic proteins of interest are technically called as transgenic or pharmed animals. For example, therapeutic proteins secreted in goat milk. There are a number of companies specializing in this technology to make products like lactoferrin, tPA, haemoglobin, melanin and interleukins in cows, goats and pigs.

Proteins started being used as pharmaceuticals in the 1920s with insulin extracted from pig pancreas. In the early 1980s, human insulin was prepared in recombinant bacteria and it is now used by all patients suffering from diabetes. Several other proteins and particularly human growth hormone are also prepared from bacteria. This success was limited by the fact that bacteria cannot synthesize complex proteins such as monoclonal antibodies or coagulation blood factors which must be matured by post translational modifications to be active or stable in vivo. These modifications include mainly folding, cleavage, subunit association, γ-carboxylation and glycosylation. They can be fully achieved only in mammalian cells, which can be cultured in fermentors at an industrial scale or used in living animals. The technology to express the desired proteins in cattle and buffaloes is yet to be developed. Again, the proteins expressed at present are of human importance. Similar approaches can also be used to produce the proteins/other products which are of importance from the dairy animal health point of view.

The effective catalytic properties of enzymes have already promoted their introduction into several industrial products and processes. Recent developments in biotechnology, particularly in areas such as protein engineering and directed evolution, have provided important tools for the efficient development of new enzymes. This has resulted in the development of enzymes with improved properties for established technical applications and in the production of new enzymes tailor-made for entirely new areas of application where enzymes have not previously been used. Enzymes are used in several different industrial products and processes. Biotechnology has enabled enzymes to be developed for processes, where no one would have expected an enzyme to be applicable just a decade ago.

Several transgenic animal species can produce recombinant proteins but presently two systems started being implemented. The first is milk from farm transgenic mammals which has been studied for 20 years and which allowed a protein, human antithrombin III, to receive the agreement from EMEA (European Agency for the Evaluation of Medicinal Products) to be put on the market in 2006. The second system is chicken egg white, which recently became more attractive after essential improvement of the methods used to generate transgenic birds. Two monoclonal antibodies and human interferon-â1a could be recovered from chicken egg white. A broad variety of recombinant proteins were produced experimentally by these systems and a few others. This includes monoclonal antibodies, vaccines, blood factors, hormones, growth factors, cytokines, enzymes, milk proteins, collagen, fibrinogen and others. Although these tools have not yet been optimized and are still being improved, a new era in the production of recombinant pharmaceutical proteins was initiated in 1987 and became a reality in 2006.

Any strategies for the commercial application of animal biotechnology must include a careful review of regulatory and social concerns. Careful review of industry infrastructure is also important.

Today we have core competencies that provide a wealth of opportunities for the members of this society, commercial companies, producers, and the general population. Successful commercialization will benefit all of the above stakeholders.

As we enter a new millennium, the research with the greatest likely impact on both the biological sciences and the biotechnology industry will be the sequencing of the human and other genomes. Widespread interest in farm animal genomics as a method for identifying genes controlling commercially important traits started only a decade ago. Although the genomics of farm animals was relatively late to arrive on the scene compared with the genomics of crop plants, it has the advantage of being able to access the enormous amount of human genome information.

One of the factors influencing the biotechnology sector's success is improved intellectual property rights legislation and enforcement worldwide, as well as strengthened demand for medical and pharmaceutical products to cope with an ageing, and ailing population all over the world.

Gene Therapy

If results of genetic engineering sound like science fiction, an even more remarkable technology called "gene therapy" is on the way. This new technology allows scientists to alter or replace genes responsible for hereditary diseases in animals and man. Gene therapy can be broadly defined as the transfer of defined genetic material to specific target cells of a patient for the ultimate purpose of preventing or altering a particular disease state. Genes and DNA are now being introduced without the use of vectors and various techniques are being used to modify the function of genes *in vivo* without gene transfer. If one adds to this the cell therapy particularly with use of genetically modified cells, the scope of gene therapy becomes much broader. Gene therapy can now combined with antisense techniques such as RNA interference (RNAi), further increasing the therapeutic applications. Gene therapy may be used for treating, or even curing, genetic and acquired diseases like cancer and AIDS by using normal genes to supplement or replace defective genes or to booster a normal function such as immunity.

In gene therapy, only somatic cells are targeted for treatment. Therefore any changes to the genes of an individual by gene therapy will only impact on the cells of their body and cannot be passed on to their children. Changes to the somatic cells cannot be passed on to future generations (inherited). Somatic gene therapy treats the individual and has no impact on future generations.

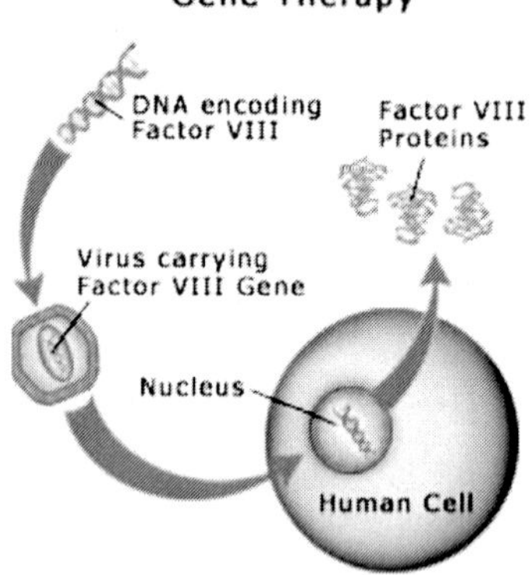

Gene therapy faces many obstacles before it can become a practical approach for treating disease. At least four of these obstacles are as follows:

High costs : Since gene therapy is relatively new and at an experimental stage, it is an expensive treatment to undertake. This explains why current studies are focused on illnesses commonly found in developed countries, where more people can afford to pay for treatment. It may take decades before developing countries can take advantage of this technology.

Limited knowledge of the functions of genes : Scientists currently know the functions of only a few genes. Hence, gene therapy can address only some genes that cause a particular disease. Worse, it is not known exactly whether genes have more than one function, which creates uncertainty as to whether replacing such genes is indeed desirable.

Multigene disorders and effect of environment : Most genetic disorders involve more than one gene. Moreover, most diseases involve the interaction of several genes and the environment. For example, many people with cancer not only inherit the disease gene for the disorder, but may have also failed to inherit specific tumor suppressor genes. Diet, exercise, smoking and other environmental factors may have also contributed to their disease.

One of the important areas of research is now **personalized medicine**, more broadly defined as **tailored therapies**. The ultimate goal is to provide the right medicine, at the right time, at the right dose, for patients. Biotechnology can help to achieve such an aim in near future.

Gene delivery tools : Genes are inserted into the body using gene carriers called vectors. The most common vectors now are viruses, which have evolved a way of encapsulating and delivering their genes to human cells in a pathogenic manner. An important problem is the problem of 'gene delivery' ie. how to get the new or replacement genes into the desired tissues. Some of the 'vectors' for the role of delivering the working copy of the gene to the target cells include using:

a) Harmless Viruses

One of the most promising methods currently being developed is the use of harmless viruses that can be used to carry genes into cells. Scientists now have the knowledge and skills to remove the virus' own genes and to replace them with working human genes. Scientists manipulate the genome of the virus by removing the disease causing genes and inserting the therapeutic genes.These altered viruses can then be used to smuggle genes into cells with great efficiency. When viruses are used in this way they are known as vectors. Some of these vectors are capable of not only carrying the gene into the cell but also of inserting the gene into the genetic makeup of the cell. Once in the right location within the cell of an affected person, the transplanted gene is 'switched on'. The transplanted gene can then issue the instructions necessary for the cell to make the protein that was previously missing or altered. However, while viruses are effective, they can introduce problems like toxicity, immune and inflammatory responses, and gene control and targeting issues. In addition, in order for gene therapy to provide permanent therapeutic effects, the introduced gene needs to be integrated within the host cell's genome. Some viral vectors affect this in a random fashion, which can introduce other problems such as disruption of an endogenous host gene.

b) Stem Cells

Another technique with potential is the use of stem cells in delivering gene therapy. Stem cells are immature cells that can differentiate or develop into cells with different functions. In this technique, stem cells are manipulated in the laboratory in order to make them accept new genes that can then change their behaviour. For example, a gene might be inserted into a stem cell that could make it better able to survive chemotherapy. This would be of assistance to those patients who could benefit from further chemotherapy following stem cell transplantation.

Types of Gene Therapy

Gene therapy may be classified into the following types:

Germ Line Gene Therapy

In the case of germ line gene therapy, germ cells, i.e., sperm or eggs, are modified by the introduction of functional genes, which are ordinarily integrated into their genomes. Therefore, the change due to therapy would be heritable and would be passed on to later generations. This new approach, theoretically, should be highly effective in counteracting genetic disorders. However, many jurisdictions prohibit this for application in human beings, at least for the present, for a variety of technical and ethical reasons.

Somatic Gene Therapy

In the case of somatic gene therapy, therapeutic genes are transferred into the somatic cells of a patient. Any modifications and effects will be restricted to the individual patient only, and will not be inherited by the patient's offspring.

4

Modern Biotechnology Issues

Environmental Biotechnology

Environmental biotechnology involves the application of biological agents, along with chemical, engineering and physical processes to maintain, protect and restore the environment. There are wide ramifications of the environmental biotechnology but we will particularly concern ourselves with two aspects important for animal health and production. First is the effect of global climate change or warming on productivity of our livestock and second the environmental aspects of pollution from livestock waste and, ways and means to ameliorate the situation using modern biotechnological tools. India is a low latitude tropical country highly vulnerable to the consequences of temperature increase. Weather variability, a major constraint on growth and welfare of India, will increase with global warming. Temperature increases in the Tibetan plateau, the Hind Kush and the Himalayas will affect the volume and timing of river flows in north India. Projected precipitation changes will also increase the variability of water availability in peninsular India. Rise in sea level will affect all coastal settlements. An Intergovernmental Panel on Climate Change (IPCC) has been constituted to look into the various issues related to global climate change.

Governments are seeking a global commitment to limit the risk of temperature increase by imposing a limit on harmful emissions (Green House Gases). It has been estimated that a 2°C goal implies a limit of 450 ppm on GHG concentration in the atmosphere, which will require at least a halving of global emissions by 2050.

Global Warming and Animal Health and Production

The threat of climate change and global warming is now recognised worldwide and some alarming manifestations of change have occurred. We all know, the earth is surrounded by a cover of gasses as atmosphere. This atmosphere allows most of the light to pass through, which reaches the surface of earth. This light from sun is absorbed by the earth surface and converts into heat energy. This heat energy is re-emitted by the surface of the earth during night. Due excessive presence of some gasses in the atmosphere, this escape of heat from earth surface is prevented, resulting in heating of earth called 'global warming'. The gasses which are responsible for causing global warming are called 'greenhouse gasses'. Carbon dioxide is one of the most important greenhouse gases. This carbon dioxide mostly comes to atmosphere as air pollution from vehicles, coal-fired power plants and other industries burning fossil fuels. Human population increase and large scale deforestation are also responsible for carbon dioxide generation. Thus, Global Warming adds energy to the Earth's biosphere.The climate change which we are experiencing is due to global warming. Heat is the fuel of weather systems. More heat, more extreme weather. Thus, it shouldn't be surprising that the result is more extreme weather. More rain, more drought and more storms.

The harmful effects of presence of greenhouse gasses in atmosphere are global warming, climate change, ozone depletion, sea level rise, adverse effects on biodiversity etc. Abnormal rise in greenhouse gas, methane, in the earth atmosphere causing arctic ice to vanish in a couple of years.

The domestic animal population has increased by 0.5 to 2.0 percent per year during the last century. One result of this population increase is that emissions from livestock have become a significant source of atmospheric methane. In fact, domestic animals currently account for about 15 percent of the annual methane emissions. Much of the world's livestock are ruminants—such as sheep, goats, camel, cattle, and buffalo—who have a unique, four-chambered stomach. In the chamber called the rumen, bacteria break down food and generate methane as a by-product. The production rate is affected by factors such as quantity and quality of feed, body weight, age, and exercise, and varies among animal species as well as among individuals of the same species. It is estimated that ruminants on low quality feeds produce more than 75 percent of the total livestock methane emissions. Fortunately, there are ways to reduce greenhouse gas emissions from livestock production through management strategies that improve production efficiency and result in lower emissions per unit of milk or meat produced. Globally, ruminant livestock produce about 80 million metric tons of methane annually, accounting for about 28% of global methane emissions from human-related activities. An adult cow may be a very small source by itself, emitting only 80 to 110 kg of methane, but with about 100 million cattle in the U.S. and 1.2 billion large ruminants in the world, ruminants are one of the largest methane sources. In the U.S., cattle emit about 5.5 million metric tons of methane per year into the atmosphere, accounting for 20% of U.S. methane emissions.

Cattle emit methane through a digestive process that is unique to ruminant animals called enteric fermentation. Since methane represents a loss of carbon from the rumen and therefore an unproductive use of dietary energy, scientists have been looking for ways to suppress its production. The most promising approach for reducing methane emissions from livestock is by improving the productivity and efficiency of livestock production. Greater efficiency of livestock production can increase profitability and be good for the environment at the same time. This general approach has been demonstrated by the dairy industry over the past several decades as milk production increased and methane emissions decreased. Nutritional and genetic improvements are mainly responsible for making modern dairy cows more productive.

Animal health may be affected by climate change in four ways: heat-related diseases and stress, extreme weather events, adaptation of animal production systems to new environments, and emergence or re-emergence of infectious diseases, especially vector-borne diseases critically dependent on environmental and climatic conditions. To face these new menaces, the need for strong and efficient Veterinary Services is irrefutable, combined with good coordination of public health services, as many emerging human diseases are zoonoses. Climate driven and other changes in landscape structure and texture, plus more general factors, may create favourable ecological niches for emerging diseases. Changes in climatic patterns and in seasonal conditions may affect disease behaviour in terms of spread pattern, diffusion range, amplification and persistence in novel habitats. Pathogen invasion may result in the emergence of novel disease complexes, presenting major challenges for the sustainability of future animal agriculture at the global level.

Livestock waste contributes to some extent to the environmental pollution in the form of sewage etc. Landfilling and incineration are popular ways to deal with biowaste but both cause negative environmental effects such as the use of valuable land and production of dangerous gases. The use of natural and genetically engineered microorganisms for domestic animals waste treatment is an emerging field to provide benefit to mankind and the environment. The structural components of cells, cellulose and hemicellulose, make biowaste very susceptible for bioproduct development and biowaste offers biotechnology an opportunity to assist in maintaining environmental quality.

Genome Mapping

A genome is the totality of genetic material in the DNA of a particular organism. Genomes differ greatly in size and sequence across different organisms. Obtaining the complete genome sequence of a microbe provides crucial information about its biology, but it is only the first step toward understanding a microbe's biological capabilities and modifying them, if needed, for agricultural purposes.

Human Genome

A knowledge of genes is an indispensable step in the understanding of biological phenomena at the molecular and cellular level. Decoding of the human genome was necessary for understanding human evolution, the causation of disease, and the interplay between the environment and heredity in defining the human condition. A project with the goal of determining the complete nucleotide sequence of the human genome was first formally proposed in 1985. In 1990 the Human Genome Project (HGP) was officially initiated in the United States with a 15-year, $3 billion plan for completing the genome sequence. The Human Genome Project (HGP) was completed in 2003. Though the HGP is finished, analyses of the data will continue for many years.

Livestock Genomics

The interest in how the genome (DNA) of any species is organized and expressed as traits in an animal has led to a new subdiscipline of genetics, called genomics. Available and rapidly improving technologies have allowed the examination of the genome of an organism as a whole, rather than one or a few genes at a time. Thus, the interest in genome research has focused on sequencing genomes of livestock and poultry to understand how various genes function and interact (functional genomics). The genomes of farm animals such as pigs, cattle and chicken are being mapped in order to identify genes controlling traits of economic importance. To help and coordinate the U.S. genome mapping efforts in cattle, sheep, swine, and poultry species, the National Research Support Project (NRSP-8) was initiated in 1993. Recently, horses and aquaculture (fish and other water animal) species were added. The primary goal of comparative mapping is to study break-points and gene evolution by evaluating the structural and functional similarities between homologous genes.Amongst the benefits of comparative genome mapping are that a gene mapped in one species is effectively mapped in all other species for which the conserved genome structures have been identified.

Cattle Genome

In the April 24, 2009 edition of the journal Science it was reported that a team of researchers led by the National Institutes of Health and the U.S. Department of Agriculture have mapped the bovine genome. The scientists found that the cattle has approximately 22,000 genes, and 80 percent of their genes are shared with humans, and they have approximately 1,000 genes they share with dogs and rodents but are not found in humans. Using this bovine "HapMap", researchers can track the differences between the breeds that affect the quality of meat and milk yields.

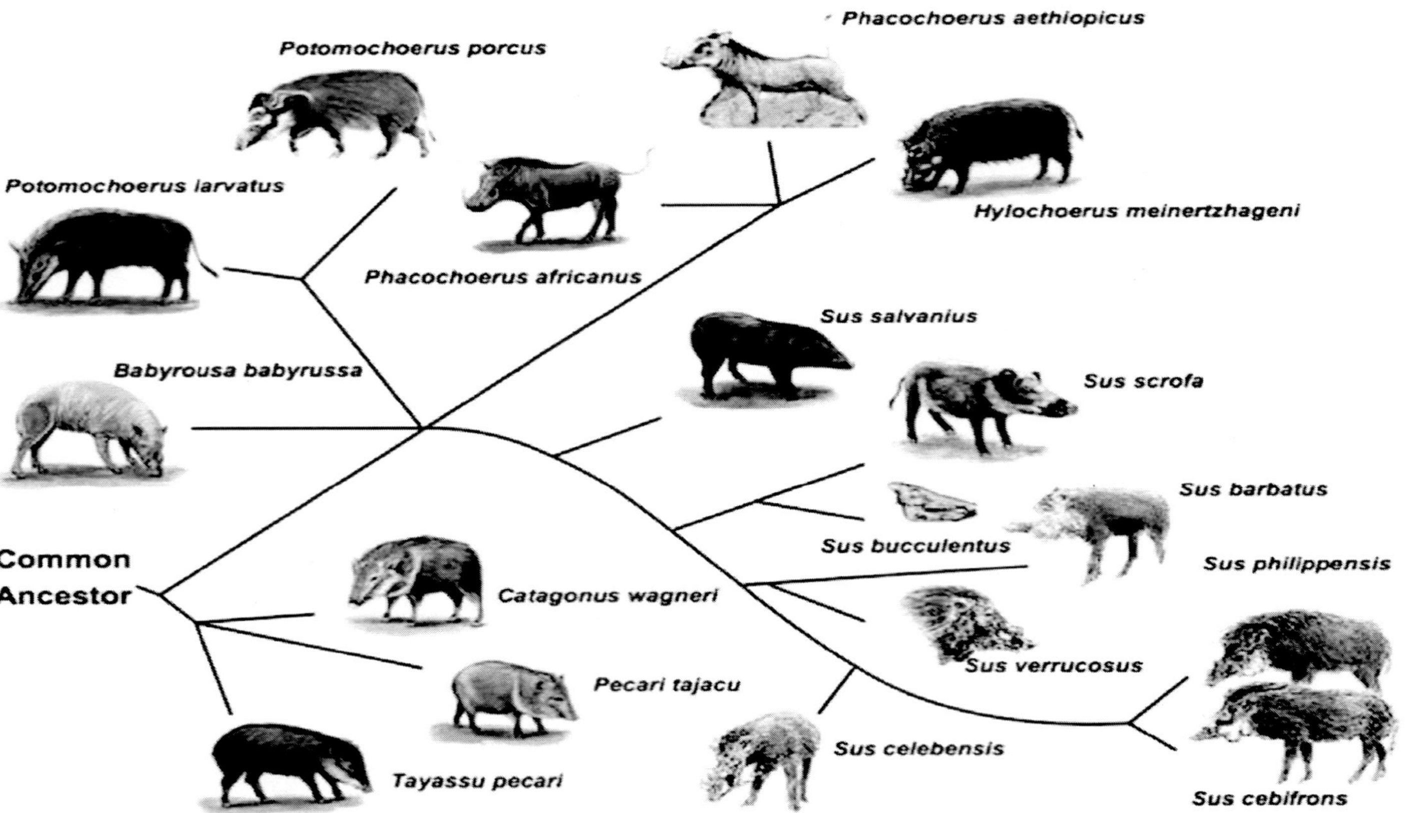

Genome Mapping and Evolutionary Genomics of the Pig

Cattle genome has 22,000 genes

The cow is the first livestock animal whose genome has been sequenced, part of an effort to read and analyze the DNA of organisms that have scientific, medical or economic importance. In addition to dozens of microbes and several plants, those sequenced so far include the chimpanzee, mouse, rat, dog, chicken, mosquito, fruit fly, opossum and platypus. Hidden in her roughly 22,000 genes are hints of how natural selection sculpted the bovine body and personality over the past 60 million years, and how man greatly enhanced the job over the past 10,000 years.

Farm animal genomics is of interest because of its use in understanding how genomics and proteomics function in various organisms. Applications such as xenotransplantation, increased livestock productivity, bioengineering new materials, products and even fabrics are several reasons for thriving farm animal genome activity and will continue to be so in times to come. Genomic information can be used to understand inherited diseases in human and domestic animals.Pharmacogenomics and personalized medicine is another area where information gleaned from genome project will be put to use. However, the current focus of attention is genomic scanning and QTL mapping in humans as well as domestic animals as this is expected to yield immediate results in terms of improvement of human and animal welfare.

Pig Genome

Pigs were among the first animals to be domesticated and pork is one of the most widely eaten meats in the world today. The past five years have seen a tremendous rise in porcine transcriptomic data. Available porcine Expressed Sequence Tags (ESTs) have expanded greatly, with over 623,000 ESTs deposited in Genbank. Early use of exotic breed crosses and now commercial breed crosses for quantitative trait loci (QTL) scans and candidate gene analyses have identified 1,675 QTL. (Int J Biol Sci 2007; 3(3) Special issue on Swine Genomics).

Bioinformatics

Bioinformatics is the application of information technology to the field of molecular biology.Computational biology is an emerging field in both for computer science and biological science, which deals with computational problems in biological fields mainly in genomics and proteomics. Computers are used to gather, store, analyze and integrate biological and genetic information, which can then be applied to gene-based drug discovery and development. Bioinformatics now entails the creation and advancement of databases, algorithms, computational and statistical techniques, and theory to solve formal and practical problems arising from the management and analysis of biological data. Most of the emerging fields of bioinformatics are related to genomics, proteomics, medical image processing, DNA microarray, protein microarray, computer aided drug designing, object oriented system programming, economic impact and new languages of bioinformatics. Common activities in bioinformatics include mapping and analyzing DNA and protein sequences, aligning different DNA and protein sequences to compare them and creating and viewing 3-D models of protein structures. Computational technologies are used to accelerate or fully automate the processing, quantification and analysis of large amounts of high-information-content biomedical imagery. Biomedical imaging is becoming more important for both diagnostics and research.

Major Branches of Bioinformatics Include

- Genomics, Proteomics,
- Computer-Aided Drug Design,
- Bio Data Bases and Data Mining,
- Molecular Phylogenetics,
- Microarray Bioinfromatics. Micro arrays are tiny chips that are used to study a phenomenon called gene expression. All the genes in all the cells are not active all the time. Which of them are "expressed" at any given time/situation is the question that microarrays help to answer.
- Systems Biology.

List of Bioinformatics Software Commonly Used in Animal Genomics Research

An introductory list of software used in animal genomics research can be found at http://www.animalgenome.org/bioinfo/resources/intro.html

Sequence Analysis Software

Blast, Blastz, Fasta, GMAP, Mulan, SynView, SynBrowse, ECR Browser

Genetics Analysis Software

CarthaGene, Crimap, JoinMap, LocusMAP, Mapmaker, NCBI CONCORDE, RHMapper, Sibs, Locusmap, SCAT, HOTSPOTTER

Functional Genomics Software

EST Scan, ORF Finder, ORFpredictor, Genome Scan, Pfam, PRINTS, PROSITE, Splice Predictor.

Biofuels

The massive energy crunch triggered by the 1973 oil price hike led to the revival of interest in non-conventional and renewable energy sources worldwide. Thus, along with the solar and the wind energy, the long neglected but potentially rich biomass became the focus of intensive utilisation for energy generation. The increase in greenhouse gas emissions and the resulting climatic changes have understandably caused worldwide concern. According to an assessment by the Intergovernmental Panel on climate change; the rise in the average temperature by the end of the next century, i.e., 2100 will be between 1 to 3.5°C. This has serious implications on the entire ecosystem. This fact has led to a series of initiatives at the international level to develop eco-friendly alternatives that would meet the needs of the present generation without compromising the abilities of future generations. This calls for urgent measures for minimizing, if not replacing, the reliance on fossil fuels to meet the increasing energy requirements. Biofuels are transportation fuels produced from biomass which means living organisms - usually plants. Petrol and diesel are actually ancient biofuels. But they are known as fossil fuels because they are made from decomposed plants and animals that have been buried in the ground for millions of years. Wood and other forms of biomass are one of the main renewable energy sources available and provide liquid, solid and gaseous fuels. Biofuels in liquid form such as fuel ethanol or biodiesel, or gaseous form such as biogas or hydrogen, are simply transportation fuels derived from biological (e.g. agricultural) sources.

- Cereals, grains, sugar crops and other starches.
- Cellulosic materials, including grasses, trees, and various waste products from crops, wood processing facilities and municipal solid waste.
- Oil-seed crops (e.g. rapeseed, soybean and sunflower) can be converted into methyl esters, a liquid fuel which can be either blended with conventional diesel fuel or burnt as pure biodiesel.
- Organic waste material can be converted into energy forms which can be used as automotive fuel: waste oil (e.g. cooking oil) into biodiesel; animal manure and organic household wastes into biogas (e.g. methane); and agricultural and forestry waste products into ethanol.

The use of biomass as fuels help to reduce the greenhouse gas emission because the CO_2 released during combustion or conversion of biomass to chemicals is that removed from the environment by photosynthesis during the production of the biomass (i.e. plant growth).

Biofuels fall into two broad categories based on their feedstock and the process used to produce the finished product.

- First-generation biofuels
- Second-generation, or "advanced," biofuels

First-generation biofuels are produced in two ways. One way is through the fermentation of either a starch-based food product, such as corn kernels, or a sugar-based food product, such as sugar cane, into ethanol, also known as ethyl alcohol, or "gasohol." Another way is by processing vegetable oils, such as soy, rapeseed and palm, into biodiesel, a nonpetroleum-based diesel fuel. Biodiesel is a clean burning fuel that is produced from renewable resources like algae, Jatropha etc. It contains no petroleum, but can be blended at any level with petroleum diesel to create a biodiesel blend. Biodiesel is already used in several countries. In Germany for example cars drive on a mixture of 60% of conventional diesel and 40% biodiesel. The experiences so far are highly positive. As yet biodiesel is too coarse for most of Jatropha - The Biodiesel Plant, the diesel engines to run on exclusively. But even if 40% of all diesels would be replaced by biodiesel, a great part of our goal would be achieved. Even the US's first commercial jet flight (Boeing 737-800) completed 2 hours test flight with one engine powered by a 50-50 blend of regular petroleum based jet fuel and a synthetic alternative made from Jatropha and algae.

Table 1: Comparison of some sources of biodiesel (*Source:* Yusuf, 2007)

Crop	Oil Yield (L/ha)	Land Area Needed (M ha) a	Percent of Existing US Cropping Area
Corn	172	1540	846
Soybean	446	549	326
Canola	1190	223	122
Jatropha	1892	140	77
Coconut	2689	99	54
Oil palm	5950	45	24
Microalgae b	136900	2	1.1
Microalgae c	58700	4.5	2.5

First-generation biofuels are not a long-term solution. There is simply not enough available farmland to provide more than about 10 percent of developed countries' liquid-fuel needs with first-generation biofuels. The additional crop demand will raise the price of animal feed and make some food items more expensive.

The most commonly used biofuel is ethanol. Bioethanol is a liquid fuel distilled from plant material and recycled elements of the food chain. It can be produced from crops such as cereals, oilseeds, sugar beet and fodder. It is produced first by fermentation, followed by distillation and finally dehydration. It is renewable and sustainable. Bioethanol is a petrol additive substitute which can be used in cars, vans, buses, lorries, agricultural vehicles, boats etc.

Advantages of Bioethanol

- Bioethanol can cut emissions of carbon dioxide, a greenhouse gas, by 50 to 60% compared to fossil fuels.
- It is generally accepted that on a well to wheel basis, bioethanol gives a 70% carbon dioxide reduction versus petrol. This means that a 5% blend produces 3.5% less carbon emissions, whilst an 85% blend would achieve a 50% reduction.
- The crops used for bioethanol are normal farm crops which can be grown using conventional farming techniques.

This is the same stuff as in alcoholic drinks, except that it's made from corn that has been heavily processed. Countries around the world are using various kinds of biofuels. For decades, Brazil has turned sugarcane into ethanol. For the future, many think a better way of making biofuels will be from grasses and saplings, which contain more cellulose. Cellulose is the tough material that makes up plants' cell walls, and most of the weight of a plant is cellulose. If cellulose can be turned into biofuel, it could be more efficient than current biofuels, and emit less carbon dioxide. For technical reasons, today's standard vehicle engines can only use fuel with small amounts of ethanol or FAME blended in (5-10%).

Second-generation, or "advanced," biofuels, made from nonfood sources, hold significant promise as a low carbon, renewable transportation fuel that can complement traditional petroleum based fuels in meeting the world's future energy needs. Research into this experimental process is focused on developing technologies that can convert cellulosic biomass, often regarded as a waste material, into transportation fuels. Examples of cellulosic biomass include:

- Agricultural wastes, such as corn stalks and husks
- Forestry wastes, such as wood chips and tree trimmings
- Fast-growing trees and grasses grown as energy crops
- Waste paper
- Food processing wastes

Although using cellulosic biomass as a source of new transportation fuels have obvious advantages, these materials have different chemical structural bonds than food-based crops and are difficult to break down, especially on a large scale. These second-generation fuels may play an important role in diversifying the world's energy sources and curbing greenhouse gas emissions. Cellulosic biofuels are liquid fuels made from inedible parts of plants which offer the most environmentally attractive and technologically feasible near-term alternative to oil.

Demand for biofuels is expected to be on the rise in time to come. Biofuels is defined as fuel such as methane produced from renewable biological resources such as plant biomass and treated municipal and industrial waste. There are different biofuels. e.g. ethanol is made from corn. The starch in the kernel of the corn is hydrolysed to sugars by enzymes or acid hydrolysis. The sugars are fermented to ethanol and ethanol concentrated and purified by distillation. Biodiesel is made of two biodegradable substances alcohol and vegetable oil. It's therefore less toxic as compared to normal diesel. Biodiesel's lack of polluting emissions is the most dramatic difference between it and petroleum. Biofuels cause reduction in emissions as compared to petroleum diesel. Biodiesel is generally available in Europe.

Even with these new improved methods, one constraint on the potential of conventional biofuels is that they use food crops. If biofuel production is to scale-up to help meet growing transport fuel demand then non-food raw materials need to be developed. This means looking at new feedstocks and new processes and fuels beyond ethanol and fatty acid methyl esters or biodiesel. For technical reasons, today's standard vehicle engines can only use fuel with small amounts (5 to 10%) of ethanol or biodiesel blended in oil.

Advantages of biofuels are that they are renewable, means their sources can be regrown, and depending on the feedstock, the processing technology and the type of fuel produced, they can offer environmental benefits such as lower carbon emissions and lower sulfur compared with conventional petroleum-based fuels.

Biomass-to-Alcohols (B2A) thermo-chemical process can utilize virtually any organic feedstock, such as woodchips, corn stover, sugar bagasse, wheat straw and so forth, to produce highly sustainable and renewable fuel. This technology is a second generation biofuel technology which has the following advantages.

Ethical use of waste material. No need for agricultural lands or dedicated energy crops.

Ecological

- Greater greenhouse gas reductions.
- Sustainable and renewable feedstocks.
- Environmental friendly production process.

Economical

- Cheaper to produce than corn or enzymatic ethanol.
- Multiple revenue streams.
- Greater flexibility.

And unlike other forms of renewable energy (like hydrogen, solar or wind), biofuels are easy for people and businesses to transition to without special apparatus or a change in vehicle.

Looking to the Future: The Potential of Second-Generation Biofuels

Many uncertainties remain for the future of biofuels, including competition from unconventional fossil fuel alternatives and concerns about environmental tradeoffs. Perhaps the biggest uncertainty is the extent to which the land intensity of current biofuel production can be reduced. The amount of biofuel that can be produced from an acre of land varies from 100 gallons per acre for EU rapeseed to 400 gallons per acre for U.S. corn and 660 gallons per acre for Brazilian sugarcane.

Cellulosic ethanol could raise per acre ethanol yields to more than 1,000 gallons, significantly reducing land requirements. Cellulosic ethanol is made by breaking down the tough cellular material that gives plants rigidity and structure and converting the resulting sugar into ethanol. Cellulose is the world's most widely available biological material, present in such low-value materials as wood chips and wood waste, fast-growing grasses, crop residues like corn stover, and municipal waste.

U.S. cellulosic fuel production costs are now estimated at more than $2.50 per gallon, compared with $1.65 per gallon for corn ethanol. Venture capital and government subsidies are supporting companies interested in making cellulosic ethanol commercially viable, primarily in the United States, but also in several other countries, including Canada, Brazil, China, Japan, and Spain.

In the meantime, other costs of cellulosic ethanol production need to be fully assessed, such as the impacts of harvesting grasses, trees, and crop residues on the erodibility and fertility of land resources. There are also questions regarding the upstream logistical and environmental costs of harvesting, transporting, and storing large volumes of bulky feedstock used in processing.

Future Role of Biofuels Depends on Profitability and New Technologies

Technological advances and efficiency gains—higher biomass yields per acre and more gallons of biofuel per ton of biomass could steadily reduce the economic cost and environmental impacts of biofuel production. Biofuel production will likely be most profitable and environmentally benign in tropical areas where growing seasons are longer, per acre biofuel yields are higher, and fuel and other input costs are lower. For example, Brazil uses bagasse, which is a byproduct from sugar production, to power ethanol distilleries, whereas the United States uses natural gas or coal.

The future of global biofuels will depend on their profitability, which depends on a number of interrelated factors. Key to this will be high oil prices: 6 years of steadily rising oil prices have provided economic support for alternative fuels, unlike previous periods when oil prices spiked and then fell rapidly, undercutting the profitability of nascent alternative fuel programs. On the other hand, the sector's profitability has been negatively affected by rising feedstock prices (corn and vegetable oil, not sugar), which account for a very large share of biofuel cost of production. For this commodity-dependent industry, government support to reduce profit uncertainty has been a common theme in the U.S., Brazil, and the EU, where biofuel production has been most significant.

Biofuels will most likely be part of a portfolio of solutions to high oil prices, including conservation and the use of other alternative fuels. The role of biofuels in global fuel supplies is likely to remain modest because of its land intensity. In the U.S., replacing all current gasoline consumption with ethanol would require more land in corn production than is presently in all agricultural production. Technology will be central to boosting the role of biofuels. If the energy of widely available, cellulose materials could be economically harnessed around the world, biofuel yields per acre could more than double, reducing land requirements significantly.

5

Information Sources and Requirements for a Biotechnology Laboratory

Information Sources for Biotechnology

Information on biotechnology applications, processes and technology is very important but difficult to find. As the field is very very active, important new information is being published almost everyday and it becomes almost impossible to keep track of all the recent information. Lot of information is secret and is available only in the patent applications or with the researchers in many private or government laboratories. However, to get the basic information several traditional sources of information like books, journals, reviews and abstracting journals are still very good. Internet search engines like Ask.com, Google.com.

- Google scholar can provide valuable information. Google Scholar (Scholar) is an index for scholarly articles in the field of biotechnology.
- http://csaweb109v.csa.com for search of Agricola data base.
- National Library of Medicine and Index Medicus, Medline and MedlinePlus

- Sciencedirect.com
- Agricola, http://agricola.nal.usda.gov/
- CAB, Library and Information Abstracts http://www.cabdirect.org/ http://www.cabi.org/animalscience/default.asp
- National Agricultural Library (NAL) of US http://www.nal.usda.gov/
- United States Department of Agriculture (USDA) http://www.usda.gov
- http://www.agnic.org/
- http://www.fao.org
- http://www.fao.org/agris/search/search.do

Most of the online journals can be available on internet provided your institutions are subscribers to those journals.

The information sources in biotechnology are similar to those in science ie abstracts and indexes, encyclopedias, dictionaries, directories, handbooks, bibliographies, and Web sites.

Books

General Biotechnology Books

A very useful list of biotechnology books for practical biotechnology techniques is available (http://www.sigmaaldrich.com/labware/labware-products.html?TablePage=14573947).

- B.R. Glick and J.J. Pasternak. Molecular Biotechnology, 3rd edn, ASM Press, 2003.
- Biotechnology: A Comprehensive Treatise, ed by H.J. Rehm and G. Reed, VCH Verleg, Germany, 1999.
- Introduction to Environmental Biotechnology by A.K. Chaterjee, PHI, India, 2000.
- Environmental Biotechnology by Alan Scragg, Pearson Education Limited, UK. 1999.
- Biotechnology : An Introduction (with InfoTrac) by Susan R. Barnum, Paperback: 336 pages, Publisher: Brooks Cole.

- Environmental Biotechnology: Principles and Applications (2001), B.E. Rittman and P.L. McCarty, McGraw Hill.
- Biotechnology: Demystifying the Concepts by David Bourgaize, Thomas R. Jewell, Rodolfo G. Buiser, Paperback: 416 pages, Publisher: Pearson Education.
- Introduction to Biotechnology by William J. Thieman, Michael A. Palladino, Paperback: 350 pages, Publisher: Benjamin Cummings.
- Basic Biotechnology by Colin Ratledge (Editor), Bjorn Kristiansen, Paperback: 584 pages, Publisher: Cambridge University Press.
- H. Robert Malinowsky, Reference Sources in Science, Engineering, Medicine, and Agriculture.
- Lehninger Principles of Biochemistry 5th Edition David L. Nelson and Michael M. Cox.
- Molecular Biology of the Gene. James D. Watson, Tania A. Baker, Stephen P. Bell, Alexander Gann, Michael Levine, Richard Losick, Inglis CSHLP 6th Edition Hardcover Benjamin Cummings.

There has been a virtual explosion of information related to the biological sciences that is available on the Internet and the World Wide Web. When a simple Internet search yields hundreds of possibilities, how can you decide quickly and effectively which resources are the best for your particular set of circumstances?

Animal Biotechnology Books

- Adams, C.E., ed. (1982). Mammalian Egg Transfer. Boca Raton, FL, CRC Press. 242 pp.
- Betteridge, K.J., ed. (1977). Embryo Transfer in Farm Animals, Monograph 16. Ottawa, Canada, Department of Agriculture. 92 pp.
- Brackett, B.G., Seidel, G.E., Jr. and Seidel, S.M., eds. 1981. New Technologies in Animal Breeding. Orlando, FL, Academic Press. 268 pp.
- Clark, A.John (1998) Animal Breeding Technology for the 21st Century Vol 4 of Modern Genetics, CRC Press pp 252.
- Daniel, J.C. Jr., ed. (1978). Methods in Mammalian Reproduction. Orlando, FL, Academic Press. 566 pp.

- Daniel, J.C., Jr. ed. (1971). Methods in Mammalian Embryology. San Francisco, CA, W.H. Freeman & Company. 532 pp.
- Donaldson, L.E., ed. (1982). Embryo Transfer in Cattle. San Antonio, Rio Vista International. 148 pp.
- Elsden, R.P. and Seidel, G.E., Jr. (1985). Procedures for recovery, bisection, freezing and transfer of bovine embryos. Fort Collins, Colorado State University. 43 pp.
- Elsden, R.P. and Seidel, G.E., Jr. (1986). Procedimientos para recolección, división, congelación y transferencia de embriones bovinos. Fort Collins, Colorado State University. 45 pp.
- Evans, J.W. and Hollaender, A., ed. (1986). Genetic Engineering of Animals. New York, Plenum Press. 328 pp.
- Gordon, Ian. R. (2004) Reproductive Technologies in Farm Animals. CABI Publications pp 332.
- Gwatkin, R.B.L., ed. (1986). Developmental Biology, Vol. 4: Manipulation of Mammalian Development. New York, Plenum Press. 388 pp.
- Hawk, H.H., ed. (1979). Animal Reproduction, Beltsville Symposia in Agricultural Research III. Montclair, NJ, Alanheld, Osmun & Co. 434 pp.
- Hafez, B. (2000) Reproduction in Farm Animals,7th Ed, Wiley Blackwell, pp 509.
- International Embryo Transfer Society. (1987). Manual. 309 W. Clark Street, Champaign, IL 61820. 87 pp.
- International Embryo Transfer Society. Proceedings of annual conferences. 1978–1989 January issues of Theriogenology. Available from publisher or IETS, 309 West Clark St., Champaign, IL 61820.
- Kahn Wolfgang. (2004) Veterinary Reproductive Ultrasonography, Schlütersche, Hannover.
- Mastroianni, L., Jr. and Biggers, J.D., ed. (1981). Fertilization and embryonic development *in vitro*. New York, Plenum Press. 371 pp.
- Pinard van der Laan M.H., Gay C., Pastoret P.P., Dodet B. Eds. (2008). Animal Genomics for Animal Health. Developments in Biologicals, Basel, Karger, Vol. 132., 439p.

- Ruvinsky, A., Marshall Graves, J.A. Ed. (2005). Mammalian Genomics. CABI Publishing Series, pp 600.
- Seidel, G.E., Jr., ed. (1985). Technology in animal agriculture for investment strategists. Fort Collins, Colorado State University. 337 pp.
- Seidel, G.E., Jr., ed. (1988). Techniques for freezing mammalian embryos. Fort Collins, Colorado State University. 90 pp.
- Srivastava, A.K., Singh, R.K. and Yadav, M.P. (2005). Animal Biotechnology, Oxford and IBH Publishing Co., New Delhi.
- Zoe Lacroix Eds. Terence Critchlow Eds. Zoi LaCroix (Editor) (2003)Bioinformatics, Elsevier.

Review Articles

Abu Nasar Md Aminoor Rahman, Ramli Bin Abdullah and Wan Embong Wan Khadijah (2008). A review of reproductive biotechnologies and their application in goat. Biotechnology 7(2) 371-384.

Amoah, E.A. and S. Gelaye. (1997). Biotechnological advances in goat reproduction. J. Anim. Sci. 75:578–585.

Betteridge K.J. (2003). A history of farm animal embryo transfer and some associated techniques. Anim. Reprod Sci. 79:203-244.

Bishop, S.C. and Morris, S.A. (2007) Genetics of disease resitance in sheep and goats. Small Rumi. Res. 70: 48-59.

Brown, C.M. (1997). Where To find nutritional science journals on the World Wide Web. J. Nutri. 127(8):1527-1532.

Brown, C.M. (2005). Where do molecular biology graduate students find information? Science and Technology Libraries 25(3): 89-104

Brys, M. (1997). Some useful hosting centers and databases for molecular biology on the World Wide Web. Acta Biochimica Polonica 43(4):743-751.

Carter, P.B. (2003). The current state of veterinary vaccines: is there hope for the future? J. Vet. Med. Ed. 30:152-154.

Church, R.B. (1989). Possibilities for genetic engineering of animals. Int. J. Anim. Sei. 4 : 1-6.

Erlich, H.A.; Gelfand, D. and Sninsky, J.J. (1991). Recent advances in the polymerase chain reaction. Science. 252: 1643-1650.

Foote, R.H. (2002). The history of artificial insemination: Selected notes and notables. J. Anim. Sci. 80:1-10.

Foote, R.H., Parks, J.E. (1993). Factors affecting preservation and fertility of bull sperm: A brief review. Reprod. Fertil. Dev. 5, 665-673.

Georges, M. (1996). Biotechnology and animal production: Advantages and problems. Proc., 64th General Session, Office, International des Epizooties, Paris 20-24 May, W.H.O.

Goel, A.K. and Agrawal, K.P. (1992). A review of pregnancy diagnosis techniques in sheep and goats Small Rumi. Res. 9:255-264.

Goel, A.K. and Kharche, S.D. (2009) Advances in reproductive technique in goats-A review. Indian J. Small Rumi. 15(1)1-34.

Hansen P.J., Block, J. (2004). Towards an embryocentric world: the current and potential uses of embryo technologies in dairy production. Reprod. Fert. Dev. 16:1-14.

Hart, Judith L. and Dart, Gary E. (1997). Biotechnology resources (Internet resources column). College and Research Libraries News 58(11):759-762.

Hasler J.F. (2003). The current status and future of commercial embryo transfer in cattle. Anim. Reprod. Sci. 79:245-264.

Hein, W.R. and Harrison, G.B.L. (2005). Vaccines against veterinary helminthes. Vet. Parasitol. 132: 217-222.

Jindal, S.K. Arvind Kumar, Hooda, O.K., Solanki, V.S. and Punj, M.L. (1998). Diagnostic ultrasound in reproductive physiology of domestic animals. Dairy Guide 25-26.

Jindal, S. K.; Sajjan Singh; Pawan Singh and Jain, G.C. (1994). A decade of embryo transfer technology in buffaloes. Problems and prospects. Indian Dairyman. 46: 373-375.

Jindal, S.K. and Madan, M.L. (1987). Future technologies for improvement in animal production system with special reference to Asia. Dairy Guide, Jan. 1987 pp 15-17.

Jindal, S.K., Mahendra Singh and Ludri, R.S. (1989). Growth hormone for more milk production. Indian Dairyman. 40(20) : 82-84.

Jindal, S.K. and Chopra, S.C. (1993). Buffalo semen freezing, Problems and perspectives. Indian Dairyman. 45: 96-100.

Kamra, D.N., Patra, A.K., Chatterjee, P.N., Kumar, R., Agarwal, N. and Chaudhary, L.C. (2008) Effect of plant extracts on methanogenesis and microbial profile of the rumen of buffalo: a brief overview. Aust. J. Exp. Agric. 48: 175-178.

Kane, M.T. (2003). A review of *in vitro* gamete maturation and embryo culture and potential impact on future animal biotechnology. Anim. Reprod. Sci. 79:171-190.

Knox, D.P., Redmond, D.L., Skerce, P.J. and New lands, G.F.J. (2001). The contribution of molecular biology to the development of vaccines against nematode and tramatode parasites of domestic animals. Vet. Para. 101: 311-315.

Kumar, A., Saini, A., Jindal, S.K. and Punj, M.L. (1998). Cryopreservation of embryos of various domestic animals. Indian Dairyman 50(5): 5-10.

Kundu, S.S., Kundu, S., Prasad, D.N. and Jakhmola, R.C. (1988) Lignin biodegradation to improve digestibility of straws. Int. J. Anim. Sci. 3: 1-16.

Lightowlers, M.W. (2003). Vaccines against cestode parasites : anti helminth vaccine that work and why. Vet. Para. 115: 83-123.

Meeusen, Els N.T., John Walker, Andrew Peters, Paul-Pierre Pastoret, and Gregers Jungersen. (2007). Current Status of Veterinary Vaccines. Clin. Microbiol. Rev. 20(3): 489–510.

Mohanty, S.S. (1988). Biotechnology in Animal Agriculture. Int. J. Anim. Sci. 3 : 101-109.

Padha, Harish (1996). Biotechnology: Its global impact and relievance to India. Current Science. 71(9): 670-676.

Pastoret, P.P., and Jones, P. (2004). Veterinary vaccines for animal and public health. Dev. Biol. (Basel). 119:15-29.

Peeters, R.M. and Im, K.S. (1994). Transgenic livestock- review Asian Aust. J. Anim. Sci. 7: 1-17.

Robinson, J.J. and McEvoy, T.G. (1993). Biotechnology the possibilities Anim. Prod. 57: 335-352.

Sanders, D.M. (1996). Molecular biology databases on the Internet. Biotechniques 21(3):438.

Seidel G.E. Jr. (2003). Sexing mammalian sperm—intertwining of commerce, technology, and biology. Anim. Reprod. Sci. 79:145-156.

Spooner, R.L. (1997). Genetics of disease resistance and the potential of genome mapping. Trop. Anim. Hlth. Prod. 29: 95S-97S.

Swindell, Simon R., Miller, R. Russell, and Myers, Garry S.A. (1996). Internet for the Molecular Biologist Vol. 3 (Current Innovations in Molecular Biology Series.) Horizon Scientific Press, New York.

Vishwanath R. (2003). Artificial insemination: the state of the art. Theriogenology 59:571-84.

Journals

There are several important biotechnology journals. A list of such journals will be very very lengthy. However, a few important ones have been listed below.

- Nature
- Science
- Journal of Molecular Biology
- Proceedings of National Academy of Sciences (US)
- Biotechnology
- Letters in Applied Microbiology
- Current Science
- Indian Journal of Biotechnology
- Indian Journal of Experimental Biology
- Journal of Biological Chemistry
- Biotechnology Letters
- Cell Biology

Animal Biotechnology Journals

- Biology of Reproduction
- Theriogenology
- Journal of Animal Science
- Journal of Dairy Science
- Animal Feed Sciences and Technology
- Livestock Production Science
- Small Ruminant Research
- Buffalo Journal

- Indian Journal of Animal Sciences
- Indian Journal of Dairy Sciences
- Indian Veterinary Journal
- International Journal of Animal Sciences
- Asian Australasian Journal of Animal Sciences
- Indian Journal of Animal Nutrition
- Animal Reproduction Science

Biotechnology Job Listings

The Scientist

Provides a listing of career opportunities and announcements related to international science jobs, including very good representation from the biological and allied medical and health sciences. Entries taken from monthly listings in the journal, The Scientist. [{http://165.123.33.33/jobs.html}]

Nature : International Science Jobs - Life Sciences

An international listing of positions available, as noted in the journal Nature. Most are for candidates with advanced degrees. [{http://www.nature.com/naturejobs/}]

Patents

Patents are part of the legal area called intellectual property that also includes trademarks, trade secrets, and copyright. A patent is essentially a grant from the government, or a regional organization, giving the inventor the right to exclude others from making the invention for a specified period; as property, patents may be bought, sold, licensed, or bequeathed. The patenting process makes technological advances known to the public so improvement of the technology may occur while at the same time protecting the inventor from others manufacturing his invention. On the flip side, during the protection period consumers may have fewer options for obtaining a commodity and may pay higher prices.

U.S. patent applications filed prior to November 29, 2000 were confidential and unless granted patent protection, the public never saw these applications. U.S. patent applications first went public on March 15, 2001 with the release of the Patent Applications database.

Currently, inventors have the option to delay the publication of their patent application for 18 months or upon issuance, whichever comes first; if the inventor does not take the option to delay, the application is published immediately.

Most countries in the world have some type of intellectual property protection, however, not all issue patents and among those that do, there are variations in the examination process and in the types of patents issued.

- The Free Web Databases and Their Content
- http://patft.uspto.gov/ The United States Patent and trademark office website.
- Esp@cenet, The European patent office website.
- Google Patents
- http://www.istl.org/09-spring/experts2.html-A Short Course on Patent Reference for Science and Technology Librarians

Requirements for Setting up a Biotechnology Laboratory

The major requirement that distinguishes biotechnology laboratory from other laboratories is the need to maintain aseptic conditions.

The wash up and sterilization of media and other equipments should be located at one end of the room and clean area for sterilize handling at the other end, and preparation, storage and incubation in between. The storage and incubation area should be near to the sterile working area.

In recent years the introduction of laminar flow cabinets has been a great boon in providing a sterile working area and allows the use of unspecialized laboratory accommodation for culture work.

For sterile work, area should be in a quiet part of the laboratory preferably in a cubicle where a laminar flow cabinet is suitably fitted. The area should be free from dust or draughts. Nothing should be stored on the laminar flow cabinet or bench used for sterile handling except most necessary equipments like holding pipettes etc. The laminar flow cabinets provide sterile air blown over the work surface. For IVM/ IVF laboratory a horizontal laminar flow is more preferable as apposed to a vertical one because of ease of handling the ovaries etc. If an air curtain is provided at the door of the cabinet, introduction of dust etc.

can be avoided. In India laminar flow are manufactured by companies like Clenzoids, Yorco etc among others. The laminar flow size should be a minimum of 300x300 or 450x450 mm with a 12 to 18 inch, square filter size.

The room should be provided some amount of temperature control and dual thermostats in parallel are standard for most IVM/IVF laboratories in the west. The door for such laboratories should be well insulated and preferably self closing.

It may be convenient to position a bench to carry microscope etc close to the sterile handling area, either dividing the area or separating it from other end of the lab. The service bench should have provisions for storage of sterile glassware, plastics, pipettes, screw caps, syringes etc., in drawer units below and shelves above. The bench may also house other accessory equipments such as a small bench top centrifuge. The bench should provide a close supply of the immediate requirements.

Media preparation : The area marked for media preparation should have a coarse and fine balance, pH meter and an osmometer. Bench space will be required for dissolving and stirring solutions, and for bottling and packaging. Heat stable solutions and equipment can be autoclaved for sterilization.

Wash-up and sterilization facilities should be in a separate room outside the tissue culture lab as the humidity and heat that they produce are difficult to dissipate. Autoclaves, ovens and distillation apparatus should be located in a separate room with an efficient exhaust fan. The wash up area should have plenty of space for soaking glassware and deep sinks for manual washing of glassware.There should be plenty of bench space for handling baskets of glassware, sorting pipettes, packaging and sealing sterile packs and a pipette washer and drier.

Storage must be provided for sterile liquids at room temperature, media at 4°C and serum, trypsin, glutamine etc. at -20°C or -70°C. Refrigerators and freezers should be located towards the nonsterile end of the laboratory. As a rough guide you will need 200 L for 4°C storage and 100 L for -20°C storage per person.

The laboratory floor should be covered with vinyl or other dust proof finish. Allow a slight fall in the level towards a floor drain located in the center of the room or on one side. This allows liberal use of water if the floor has to be washed, but more important, it protects equipment from damaging floods if stills, autoclaves or sinks overflow.

If the tissue culture lab and preparation, wash up and sterilization areas can be separated, so much the better. If you have a separate wash-up and sterilization facility, it will be convenient to have this on the same floor and adjacent.

Make sure that your laboratory doors are wide enough to allow entry to all the equipment you want, particularly laminar flow units, cold handling cabinets and ultra low temperature freezers. Also allow space for maintenance.

Equipments needed for such a laboratory can be grouped in the category of essential, beneficial and useful.

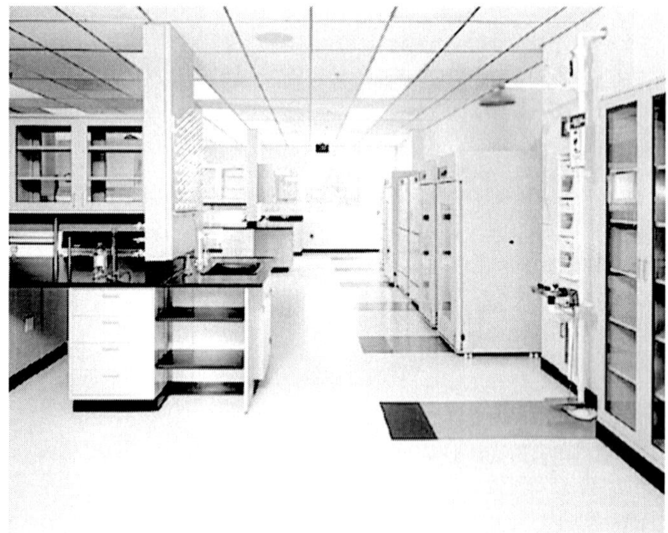

Essential

- Incubator
- Autoclave
- Oven
- Refrigerator
- Freezer
- Inverted microscope – A good microscope is most essential for any animal biotechnology laboratory. Good microscopes are available from Nikon, Wild, Zeiss, Olympus, Bausch and Lomb etc. The microscope having good optics is a good microscope and presently German, Japanese microscopes are considered to

be the best. Look for manufacturers catalog for selecting the best one suited for your specific needs.

- Soaking bath or sink
- Deep washing sink
- Pippette washer
- Water purifier or water distillation apparatus. A glass distillation apparatus was a permanent fixture in any laboratory worth the name. Now a days Millipore water purification systems or equivalent systems are commonly used for providing quality water for culture and other work. A good water purification system is essential for any biotechnological work and quality of the purifier should not be compromised. Modern purifiers works on the combined capabilities of a deionizer, reverse osmosis, Millipore or solid state filtration besides a norit or activated charcoal cartridge for providing extra pure water.
- Bench centrifuge
- Liquid nitrogen freezer
- Liquid nitrogen storage cylinders- The liquid nitrogen containers or cryocans are in fact double walled Dewar flasks. Now a day's several indigenous and foreign manufacturers provide quality cryocans. Indian manufacturer for cryocan is IBP and there are a number of good foreign manufacturers like MVE, Staborne, Air Liquide France etc.

Beneficial

- *Laminar Flow Hood (Horizontal)* : The laminar flow hood is essential for maintenance of aseptic conditions. Both vertical and horizontal models are available. Now a days Indian made laminar flow hoods are available like from Klenzoids, Yorco etc
- Cell counter
- Vacuum pump
- *CO_2 incubator* : This Instrument is one of the most essential for IVM/IVF laboratory. Initially researchers used to use a dessicator with a connection to CO_2 cylinder. Now a days excellent CO_2 incubators are available. Some prominent brands include Nuaire, Forma Scientific etc.
- Coarse and fine balance

- pH meter -
- Osmometer
- Phase contrast microscope
- Portable temperature recorder
- Magnetic stirrer
- Pipette dryer
- Automatic pipettes or dispensers
- Sterilizing oven and drying oven (separate)

Useful

- -70^0 C freezer
- Closed circuit TV
- Colony counter
- High capacity centrifuge
- Cell sizer or cell sorter
- Time-lapse cinemicrographic equipment
- Interference contrast microscope
- Polythene bag sealer
- Controlled-rate cooler
- Densitometer
- Pipetter puller
- Microforge
- DNA injector
- Micro-manipulator
- Computer

PCR

The Thermal Cycler (also known as a Thermocycler, PCR Machine or DNA Amplifier) is a laboratory apparatus used to amplify segments of DNA via the polymerase chain reaction (PCR) process. Modern PCR machines rapidly changes temperatures for PCR reactions, allowing PCR to cycle between primer annealing, DNA amplification, and strand melting cycles. Some thermal cyclers are equipped with multiple blocks allowing several different PCR reactions to be carried out simultaneously. Also some apparatus have a gradient function, which allows different temperatures in different parts of the block. This is particularly useful when testing suitable annealing temperatures for primers.

Specifications of a Good PCR Macine

Sample capacity: 96x0.2ml or 1 PCR plate

Multiple annealing temperature capability in a single block using either gradient feature or other features for PCR optimization, ranging from 1°C to at least 20°C across the block.Temperature block homogeneity should be = 0.3°C for annealing and extension steps. Temperature control accuracy should be ± 0.25°C. Ramp time should be minimum 4°C/sec for heating and 3°C/sec for cooling. Automatic restart after power failure. Upgradable to real-time PCR. Any feature that actively minimizes sample evaporation during run. Ease of programming and program data archival to external media.

Some of the prominent manufacturers of PCR

- Stratagene's RobotCycler
- MJ research Peltier Thermal Cycler
- Perkin Elmer DNA Thermal cycler
- BioRad.

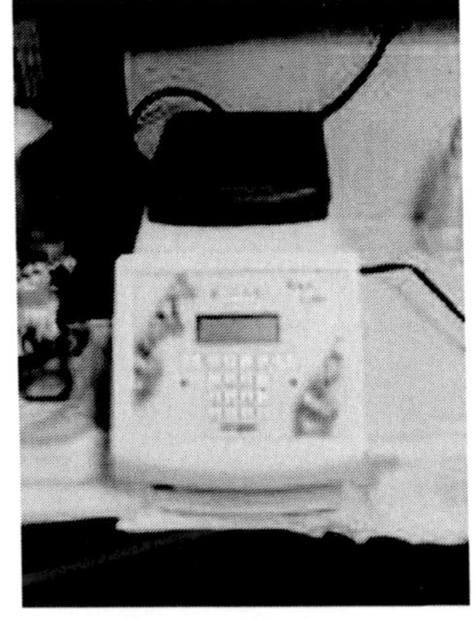

Gel Documentation System

Electrophoresis

Water Purification : Water is purified for two main purposes. (1) rinsing glassware and (2) making up media and reagent. For the first, a simple deionizer or reverse osmosis unit is adequate. For the second, higher purity water is required. The traditional method was to use double glass distilled water (glass or silica or quartz sheathed elements) but many laboratories now replace the first distillation stage with deionization. The deionizer should have a conductivity meter monitoring the effluent to signify when the cartridge must be changed and quality of water output. Deionization alone is insufficient, as many organic contaminants are not retained by the ion-exchange resins. Distillation is usually recommended as a necessary second step.

Now a days, high grade water filtration systems have become available especially Millipore etc. These filtration systems use multiple filtration techniques including high-grade deionization, charcoal or norit filtration, reverse osmosis, ion exchange resin filtration and finally through solid state Millipore filters. Although expensive, this has now become standard for most IVM/IVF laboratories the world over.

Personal Hygiene, while washing etc has been much debated but use of aprons, surgical gloves, caps, gowns and face masks are often recommended. Some of them may be absolutely necessary but others do sometime help. If you have long hair, tie it back (especially for ladies). Do not talk while working asceptically. If you have a cold do not do any tissue culture during the infection.

Standard glass or disposable plastic pipettes are easiest form of manipulating liquids. Syringes are often used but not recommended as they produce high shearing forces when dispensing cells and increase the risk of self inoculation. Deep screw caps should be used in preference to stoppers and when washing caps it must be ensured that all detergent is rinsed from behind rubber liners. The screw cap should be covered with aluminum foil to protect the neck of the bottle from sedimentary dust.

In a horizontal type of laminar flow, the air flow blows from the side facing you, parallel to the work surface and is not recirculated.In a vertical cabinet, the air blows down from the top of the cabinet on to the work surface and is either recirculated or vented.

If potentially hazardous material (radioisotopes) is handled, a class II vertical flow biohazard hood should be used.

If known human pathogens are handled, a class III pathogen cabinet with a pathogen trap on the vent is obligatory. Ultraviolet lights are used to sterilize the air and exposed work surfaces in laminar flow cabinets between uses.

Laboratory Safety and Hazards

No one should ignore potential hazards while working in a laboratory. Fire is one of the most important hazard. Avalability of fire safety measures should be one of the prime concern while working in a biotechnology laboratory. The laboratory must be fitted with one of the smoke detecting appliances connected to the local fire station or siren. Fire extinguishers should be available either in the laboratory or near by. The laboratory must be cross ventilated or should have appropriate number of exhaust fans.

The most common form of injury in tissue culture lab results from accidental handling of broken glass. Take care when fitting a bulb or pipetting device onto a pipette. A fire extinguisher should be wall mounted on some convenient place in the laboratory. Use of Bunsen burners should be done with care. Radiation hazards can be kept minimum, if adequate precautions are taken while handling radioisotopes. UV light sterilization should be done with adequate protection to the skin and eyes.

6

Present Status and Future Prospects of Biotechnology in Animal Health and Production

Biotechnology has potential applications in the management of several animal diseases such as foot and mouth disease, classical swine fever, avian flu and bovine spongiform encephalopathy. The most important biotechnology based products consist of vaccines, particularly genetically engineered or DNA vaccines. Gene therapy for diseases of pet animals is a fast developing area because many of the technologies used in clinical trials of humans were developed in animals and many of the diseases of cats and dogs are similar to those in humans. RNA interference technology is now being applied for research in veterinary medicine

Molecular diagnosis is assuming an important place in veterinary practice. Polymerase chain reaction and its modifications are considered to be important. Fluorescent in situ hybridization and enzyme-linked immunosorbent assays are also widely used. Newer biochip based technologies and biosensors are also finding their way in veterinary diagnostics.

Transgenic technologies are used for improving milk production and the meat in farm animals as well as for creating models of human diseases. Transgenic animals are used for the production of proteins for human medical use. Biotechnology is applied to facilitate xenotransplantation from animals to humans. Genetic engineering is done in farm animals and nuclear transfer technology has become an important and preferred method for cloning animals.

Some of the key companies engaged in veterinary biotechnology include.

- Alpharma Animal Health
- Adisseo France S.A.S.
- BASF International
- Bayer Healthcare
- Boehringer Ingelheim GmbH
- DSM Nutritional Products, Inc.
- Elanco Animal Health
- Fort Dodge Animal Health
- Heska Corporation
- Intervet Schering Plough Animal Health
- Merial Animal Health Ltd.
- Monsanto
- Novartis Animal Health Inc.
- Virbac SA
- Pfizer Animal Health

The world market for animal health products was worth of US$ 17.4 billion in 2005. More than 85% of global animal health sales are generated in 15 major markets. The US is the dominant market in the sector, generating 36% of the entire global total. No other national market is responsible for a share of more than 7%.

The structure of individual markets varies widely, reflecting a combination of factors such as climate, the prevalence of particular animal diseases and the relative importance of individual species to national livestock agriculture. Products for use in companion animals are responsible for over half of all sales in some developed markets such as the US and UK, but generate less than 5% of sales in emerging markets such as China and India, for example. (www.piribo.com/publications/animal_veterinary/world_animal_health_markets.html)

Global market growth is forecast at a compound annual rate of 4.5% during the second half of this decade, driving sales up to almost of US$ 21.7 billion in 2010. Sales in China will rise at a CAGR of 8% during the forecast period, while market value in Brazil will increase by 6% a year. Most developed markets will be considerably more subdued, however, with growth in Japan, Australia and most major European countries forecast at rates well below the global average.

Within the last decade, a new veterinary devices and diagnostics industry has started to emerge. Although still a young industry that is only just beginning to find its own identity, it is an area of veterinary medicine that has considerable growth potential. The use of diagnostic tests and equipment is now taught in many veterinary schools and discussed at conferences, and is particularly popular among younger vets - a good omen for the future.

Checks and Balances in Biotechnology Applications

As with anything new or different and potentially dangerous, there must be checks and balances. There are individuals who have been opposing the use of biotechnology. There is a fear of creating a monster of some kind that cannot be controlled. It is feared that once scientists begin tinkering with genes, especially that of humans, it may be difficult to know when to stop. There do exist, a large comprehensibility gap between the scientists, administrators and the public. Many countries of the world have strict guidelines for conducting biotechnological research. Risk assessment studies have been made mandatory before release of genetically engineered organisms. For example introduction of Bt cotton in India elicited a fierce debate at various platforms. However, most experts feel that genetic engineering technique is a powerful, safe means for modifying organisms that pose no greater risk than selective breeding and other methods of altering organisms.

Biosafety Issues in Animal Biotechnology Application

Application of GM technology for commercial animal production is faced with a number of issues and challenges the most important being safety of environment, and human and animal health. These concerns are based on the argument that recombinant DNA based GM technology differs from traditional breeding in that totally new genes using potentially risky technology are transferred between widely unrelated organisms, and the location of these genes on the recipient genome is random, unlike when-gene transfer takes place through conventional breeding. These differences demand that adequate-laboratory safeguards are used and animals developed by GM technology are rigorously assessed for their performance as also for the likely risks they pose.

Six organizations are directly or indirectly involved in international biosafety regulation:

(a) *The Convention on Biological Diversity (CBD)-1992*: Concerns conservation, sustainable use and fair and equitable sharing of benefits arising from the use of biological resources. The Cartagena Protocol on Biosafety regulates transboundary movement of living modified organisms (LMOs).

(b) *The World Trade Organization (WTO)-1995*: Deals with trade in goods and services and dispute settlement. The WTO Agreement on Application of Sanitary and Phytosanitary (SPS) Measures is related to procedures of risk analysis of plant and animal pests and diseases, and food safety.

(c) *The International Plant Protection Convention (IPPC)-1952*: Develops International Standards on Phytosanitary Measures (ISPMs) against pests of plants and plant products including GMOs.

(d) *The Codex Alimentarius Commission (CAC)-1972*: Develops international standards including-those for food safety and food labeling.

(e) *The World Organization for Animal Health (OIE)-1924*: Harmonizes trade regulations for-animals and animal products, and develops standards on animal health including for infectious animal diseases.

(f) *The Organization for Economic Cooperation and Development (OECD)-1961*: Undertakes harmonization of international regulations, standards and policies.

World Organization for Animal Health (OIE)

The OIE ensures transparency in the global animal disease situation to improve the legal framework and resources of national veterinary services. It establishes standards, guidelines and recommendations relevant to animal diseases and zoonoses in accordance with its statutes as defined in the WTO-SPS Agreement (OIE, 2006; Sendashonga *et al.*, 2005).

Environment Protection Act (1986) and
Environment (Protection) Rules (1986)

The Act relates to the protection and improvement of environment and the prevention of hazards to human beings, other living creatures, plants and property. The Act mainly covers the rules to regulate environmental pollution and the prevention, control, and abatement of environmental pollution.

The Environment (Protection) Rules cover management and handling of hazardous wastes, manufacture, storage and import of hazardous chemicals and rules for the manufacture, use, import, export and storage of hazardous microorganisms, genetically engineered organisms or cells.

36 *Biosafety Regulations of Asia-Pacific Countries*

Rules for the Manufacture, Use/Import/Export and Storage of

Hazardous Microorganisms/Genetically Engineered Organisms

or

Cells. (notified under the EP Act, 1986) (1989)

These Rules include the rules for pharmaceuticals, transit and contained use of genetically engineered organisms, microorganisms and cells and substances/products and food stuffs of which such cells, organisms or tissues form a part, LMOs for intentional introduction into the environment, handling, transport, packaging and identification.

These rules are applicable to the manufacture, import and storage of microorganisms and gene technology products.

The rules are specifically applicable to:

(a) Sale, storage and handling;

(b) Exportation and importation of genetically engineered cells or organisms;

(c) Production, manufacturing, processing, storage, import, drawing off, packaging and repackaging of genetically engineered products that make use of genetically engineered microorganisms in any way.

Recombinant DNA Safety Guidelines (1990)

The Guidelines prescribe safety measures for research, field cultivation and also the environmental impact during field applications of genetically altered material products. They are applicable to research involving genetically engineered organisms originating from genetic transformation of green plants, rDNA technology in vaccine development, and also large scale production and deliberate/accidental release of organisms, plants, animals and products derived by rDNA technology into the environment.

The Guidelines also prescribe the criteria for ecological assessment on a case-by-case basis for planned introduction of rDNA organism into the environment.

Foreign Trade (Development and Regulation) Act, 1992 (2006) (Draft Amendment)

The Act provides for the development and regulation of foreign trade by facilitating imports into and augmenting exports from India and for matters connected with it. In 2006, the government made draft amendment in the foreign trade policy, making labeling of imported GM products mandatory.

Revised Guidelines for Research in Transgenic Plants and Guidelines for Toxicity and Allergenicity Evaluation of Transgenic Seeds, Plants and Plant Parts (1998)

The Guidelines cover rDNA research on plants including the development of transgenic plants and their growth in soil for molecular and field evaluation. It also includes LMOs for contained use and intentional introduction into the environment, and LMOs for use as food or feed or for processing, pharmaceuticals and transboundary movement. The Guidelines also specify requirements for import and shipment of GM plants for research use only.

Guidelines for Generating Preclinical and Clinical Data for rDNA Vaccines, Diagnostics and other Biologicals (1999)

The Guidelines cover preclinical and clinical evaluations of rDNA vaccines, diagnostics and other biologicals/pharmaceuticals. The objectives of the preclinical studies are to define physiological, toxicological and efficacious potential of r-DNA products prior to initiation of human studies.

Both *in vitro* and *in vivo* studies can contribute to evaluating the effects of r-DNA products.

The Guidelines also cover safety, purity, potency and effectiveness of the rDNA products, *in vitro* diagnostic recombinant reagents and monoclonal antibodies, and describe in detail procedures for generating monoclonal antibodies. Sensitivity and specificity required for diagnostics of infections of widespread diseases like HIV-I/II are also prescribed.

The Food Safety and Standards Act (2006)

The objective of the Act is to bring out a single statute relating to food and to provide for a systematic and scientific development of food processing industry. The Act incorporates the salient provisions of the Prevention of Food Adulteration Act, 1954 (37 of 1954) and is based on international legislations, instrumentalities and Codex Alimentarius Commission Guidelines. The Act is in tune with the international trend towards modernization and convergence of regulations of food standards with the elimination of multi-level and multi-departmental control.

The emphasis is on (a) responsibility with manufacturers, (b) recall, (c) GM and functional foods, (d) emergency control, (e) risk analysis and communication and (f) food safety and good.

Transgenic Animals Bioreactors

The production of recombinant proteins is one of the major successes of biotechnology. Animal cells are required to synthesize proteins with the appropriate post-translational modifications. Transgenic animals are being used for this purpose. Milk, egg white, blood, urine, seminal plasma and silk worm cocoon from transgenic animals are candidates to be the source of recombinant proteins at an industrial scale.

Although the several recombinant protein produced by transgenic animals are expected to be in the market in not too distant future, a certain number of technical problems remain to be solved before the various systems are optimized. Although the generation of transgenic farm animals has become recently easier mainly with the technique of animal cloning using transfected somatic cells as nuclear donor, this point remains a limitation as far as cost is concerned. Numerous experiments carried out for the last 15 years have shown that the expression of the transgene is predictable only to a limited extent. This is clearly due to the fact that the expression vectors are not constructed in an appropriate manner. This undoubtedly comes from the fact that all the signals contained in genes have not yet been identified. Gene constructions thus result sometime in poorly functional expression vectors. One possibility consists in using long genomic DNA fragments contained in YAC or BAC vectors. The other relies on the identification of the major important elements required to obtain a satisfactory transgene expression. These elements include essentially gene insulators, chromatin openers, matrix attached regions, enhancers and introns. A certain number of proteins having complex structures (formed by several subunits, being glycosylated, cleaved, carboxylated) have been obtained at levels sufficient for an industrial exploitation. In other cases, the mammary cellular machinery seems insufficient to promote all the post-translational modifications. The addition of genes coding for enzymes involved in protein maturation has been envisaged and successfully performed in one case. Furin gene expressed specifically in the mammary gland proved to able to cleave native human protein C with good efficiency. In a certain number of cases, the recombinant proteins produced in milk have deleterious effects on the mammary gland function or in the animals themselves. This comes independently from ectopic expression of the transgenes and from the transfer of the recombinant proteins from milk to blood. One possibility to eliminate or reduce these side effects may be to use systems inducible by an exogenous molecule such as tetracycline allowing the transgene to be expressed only during lactation and strictly in the mammary gland. The purification of recombinant proteins from milk is generally not particularly difficult. This may not be the case, however, when the endogenous proteins such as serum albumin or antibodies are abundantly present in milk. This problem may be still more crucial if proteins are produced in blood. Among the biological contaminants potentially present in the recombinant proteins prepared from transgenic animals, prions are certainly those raising the major concern. The selection of animals chosen to generate transgenics on one hand and the elimination of the potentially contaminated animals, thanks to recently defined quite sensitive tests may reduce the risk to an extremely low level. The available techniques to produce pharmaceutical proteins in milk can be used as well to optimize milk composition of farm animals, to add nutriceuticals in milk and potentially to reduce or even eliminate some mammary infectious diseases.

Recently in 2009, NanoViricides, Inc., announced that the herpes simplex viral load was reduced by 99.99% or 10,000 fold in *in vitro* studies by nanoviricides(TM) drug candidates. The studies were performed by Thevac in Baton Rouge, LA, in collaboration with the Division of Biotechnology and Molecular Medicine at the LSU School of Veterinary Medicine under the supervision of Dr. Gus Kousoulas. Nanoviricides act by a novel and distinctly different mechanism compared to existing drugs. Nanoviricides are designed to mimic the human cell surface to which the virus binds.

NanoViricides, Inc. Says Flu-Cide Drug Designed to Destroy all influenza A viruses including Swine and BirdFlu. Already shown to be effective against diverse influenza subtypes such as H1N1 and different clades of H5N1. The Company has previously announced excellent results in both animal studies and cell culture studies against widely different influenza subtypes and strains. If these results are confirmed in further animal and human studies, then FluCide would likely be considered the best ever drug effective against all influenzas. The current swine flu outbreak is significant in that the H1N1 virus causing it is novel (http://www.cidrap.umn.edu/cidrap/content/influenza/panflu/news/apr2109swine.html).

First Patent in Biotechnology

In 1973-1974 Stanley N. Cohen of Stanford and Herbert W. Boyer of the University of California, San Francisco, developed a laboratory process for joining and replicating DNA from different species. In 1974 Stanford and UC applied for a patent on the recombinant DNA process; the U.S. Patent Office granted it in 1980. Some believe that patent is a turning point in the commercialization of molecular biology and a harbinger of the social and ethical issues associated with biotechnology today.

Biotechnology - Ethical Concerns

The use of biotechnology has also raised various ethical concerns.The first is whether anything theoretically could go wrong with any of the technologies. For example, is it theoretically possible that a DNA sequence from a vector used for gene transfer could escape and unintentionally become integrated into the DNA of another organism and thereby create a hazard? The second is whether the food and other products of animal biotechnology, whether genetically engineered, or from clones, are substantially different from those derived by more traditional, extant technologies. A third major concern is whether the technologies result in novel environmental hazards. The fourth concern is whether the technologies raise animal health and welfare issues. Finally, there is concern as to whether ethical and policy aspects of this emerging technology have been adequately addressed. These and similar issues are being hotly debated by various sections of the stakeholders and has many a questions unanswered. Only the future can answer some or all of the above questions. Although, religion and ethics would play an important role in determining the acceptability of biotechnology by the public especially when gene manipulation in animals and man is at issue. It is important to point out that biotechnology can be used to produce some very valuable animals that are highly productive. Only a small number of these animals would be adequate for one's need and number of many non productive animals can be reduced (Mohanty, 1988).

Some Priority Areas in the Field of Biotechnology

- Development and improvements of prophylactics, therapeutics and diagnostics for the major diseases of cattle, buffaloes, sheep and goat pox, clostridial diseases of sheep and goat, Mycoplasma and Avian Diseases (Ranikhet Disease, IBO, EDS and Marek's Diseases), through modern biotechnological methods.
- Application of specific aspects of embryo transfer technology such
- as endocrinological studies, *in vitro* fertilization and embryo
- Cryopreservation for improving buffalo, sheep and goat productivity.
- Development of transgenic animals for higher productivity, disease and stress resistance.
- Identification, development and improvement of livestock by-products, viz. sera, hormones, enzymes, proteins, etc.
- Conservation of endangered species by cryogenic storage of semen, ova or embryos.
- Novel biomaterials and products for surgical use.

Suggestions for Future Development of Biotechnology Activities in the Field of Animal Health and Production in India

Sophisticated technology and costly equipment are required besides requirement of costly consumables. Hence, laboratories may be identified for the development of biotechnology activities, especially those who have developed basic infrastructural facilities and have trained manpower so that there could be an immediate take off.

There has been a concern that economic interests in the biotechnology area could, particularly through patenting, have a constricting influence on scientific research. Several specialists are of the view that such patent claims present severe restrictions on 1) the free flow of scientific information and, (2) the activity of scientific researchers.

Epilogue

We are on the brink of a biotechnological revolution in animal health and production that would greatly improve or even completely change animal agriculture in regard to diagnosis, disease resistance and productivity. However, the application of biotechnology research to achieve these application aims is tedious, complex, time consuming and expensive. Safety associated with biotech research and products has also triggered public concern. Conscientious risk assessment need to be performed before biotech based products and applications like proteins available from gene farming, recombinant vaccines, monoclonal antibodies, gene therapy and diagnostics are put to use. Although biotechnology research requires great input, the benefit to animal agriculture is obvious and immediate. However, there are some major challenges facing animal producers intending to incorporate biotechnology based methods. Public distrust, lack of or inadequate financial support and questionable cost effectiveness, are some of these challenges. The three factors: government policy uncertainty, consumer aversion and brand risk, and access to intellectual property are major obstacles in acceptance of biotechnological modified products.

7

Biotechnology Research in India

India is one of the first few countries, among the developing countries, to have recognized the importance of biotechnology as a tool to advance growth of agricultural and health sectors as early as in 1980s. India's Sixth Five Year Plan (1980-85) was the first policy document to cover biotechnology development in the country. At the top, an apex official agency viz. National Biotechnology Board (NBTB) was set up in 1982, to spearhead development of biotechnology. The NBTB was chaired by Member (Science) of the Planning Commission and had representation of almost all the S and T agencies in the country viz.Department of Science and Technology (DST), Council for Scientific and Industrial Research (CSIR), Indian Council of Agricultural Research (ICAR), Indian Council for Medical Research (ICMR), Department of Atomic Energy (DAE) and the University Grants Commission (UGC). The National Biotechnology Board issued the" Long Term Plan in Biotechnology for India" in April 1983. This document spelt out priorities for biotechnology in India in view of the national objectives such as self sufficiency in food, clothing and housing, adequate health and hygiene, provision of adequate energy and transportation, protection of environment, gainful employment, industrial growth and balance in international trade. Later in 1986, National Biotechnology Board was upgraded to a full-fledged government department called Department of Biotechnology.

India, today, holds a small share of the global biotech market, but has all the capabilities to become a dominant player. India being the second largest food producer, offers a huge market for biotechnology products, especially agribiotech products. An estimation by CII shows that the Agri-Biotech would see growth rates of as much as 60%, Diagnostic and Therapeutics of about 25% and Vaccines of about 15%. India has a sound and widely acknowledged framework of bio-safety guidelines to deal with evaluation, monitoring and release of genetically engineered organisms and there are more than 106 institutional bio-safety committees.

After becoming an IT giant, India is now shifting its focus to the most promising industry of the future, Biotechnology. With its large pool of scientific talent, world class informationtechnology industry, and vibrant pharmaceutical sector, India is well positioned to emerge as a significant player in the global biotech arena. Combined with rising public interest in this sector, growing investment by traditional business houses, tax incentives and the significant foreign investment available, the Indian biotechnology sector is-poised to emerge as a significant force on the global biotech map. India has one of the largest agriculture sectors in the world, and varied climatic zones that can help in research and development of different agribiotech products applicable worldwide. Biotechnology research in India is carried out by the following organizations besides a host of private companies in their research laboratories.

- Indian Council of Agricultural Research (http://icar.org.in)
- Indian Council of Medical Research
- Council of Scientific and Industrial Research
- Defence Research and Development Organization
- Department of Biotechnology (Department of Biotechnology: http://dbtindia.nic.in)
- Department of Atomic Energy
- University Grants Commission

Out of this, DBT is the only agency completely devoted to R&D in biotechnology.

Animal Science Biotechnology Research Centres

In India, most of the biotechnology research on animal health and production is carried out under the aegis of Indian Council of Agricultural Research, New Delhi. The Indian Council of Agricultural Research (ICAR) is an autonomous organisation under the Department of Agricultural Research and Education, Ministry of Agriculture, Government of India. The Council is the apex body for coordinating, guiding and managing research and education in agriculture including horticulture, fisheries and animal sciences in the entire country. With over 90 ICAR institutes and 45 agricultural universities spread across the country this is one of the largest national agricultural systems in the world. ICAR is an important part of the Ministry of Agriculture, Government of India and is nodal agency for controlling all research activities in the field of Animal Health and Production besides agriculture. Achievement include : A unique National facility, High Security Animal Disease-Laboratory with P-4 measures established that played a pivotal role in providing diagnostics services for avian inluenza in the country besides developing vaccine using indigenous strains. The 80% of 140 indigenous breeds of livestock and poultry characterized phenotypically and genetically. Five breeds of indigenous livestock and poultry were conserved and charecterized both phenotypically and using molecular markers. Vrindavani breed of cattle developed with production potential of 3,500 kg milk per lactation. Graded Murrah buffaloes with 2,200 kg milk yield per lactation evolved. Improved strains of sheep for fine wool (Bharat Merino), carpet wool (Chokla, Marwari, Magra) and meat (Malpura, Nellore, Mandya, Madras Red) developed. Artificial insemination method standardized in mithun, yak, camel, goats, pig and equines; first mithun calf born through artificial insemination in India; crystoscope device developed to detect accurate time for insemination in cattle and-buffaloes. For promoting backyard poultry an early maturing poultry strain, CARI Nirbhik, producing 223 eggs by 72 weeks, developed. Hormonal modulation protocols developed to increase egg production in poultry. A new fungus genus *Cyllamyces icaris* with better fibre degrading ability identified for the first time in Indian cattle and buffaloes. Area specific mineral supplement for livestock developed. Diagnostic kits developed for detecting early pregnancy in equines Embryo transfer technology standardized in buffaloes, sheep, cattle, goat and yak. Developed indigenous diagnostics and control kits for foot and mouth disease (FMD) as import substitution. Developed live attenuated vaccine for sheep and goat plague (PPR), a pentavalent vaccine for controlling bluetongue disease in small ruminants. Developed bactriocin based preparation for effective treatment of bovine mastitis; developed an indigenous medicine, M-cure for treatment of skin disease in camel. Established serum bank facility, first of its kind in India, maintaining over 170,000 serum-samples for long-term national surveys in infectious bovine rhinotrachitis, brucellosis, rinderpest and bluetongue.

Current Status and Some of the Achievements in Animal Heatlh

In India, under biotechnology activities, efforts for the improvement of animal productivity, development of newer animal vaccines and diagnostics, molecular characterization of indigenous breeds of livestock and development of animal by-products are being made. Programmes are underway in the area of animal nutrition and development of newer animal vaccines. A novel and potent anthrax vaccine, which includes mutants of legal factor and edema factor has been developed which provides better efficacy *in vivo.* An attenuated buffalo pox virus vaccine has been developed and its field trial is underway. Vaccines for Rabies, Clostridium, Hemorrhagic septicemia, Foot and Mouth disease, Bovine brucellosis, Bovine tuberculosis etc. are in various stages of development. Phage display technique has been used as an alternative to hybridoma to produce mono specific antibodies against recombinant gag antigen of Bovine Immunodeficiency Virus. Diagnostics for paste des petites virus and buffalo pox virus have been developed and validated successfully.

A RT-PCR assay was standardized for specific detection of Border disease virus and a nested PCR has been developed for differentiation of Border disease virus, Bovine viral diarrhea virus 1 and 2. Multicentric programme on Buffalo Genomics for identification of genes of economic importance is being carried out. Structural and functional aspects of 3D scaffold of bovine origin for cardio myocyte culture are being studied. Effects are also on to develop biomaterial of bovine origin for reconstruction surgery in animals. Notable achievements in the field of animal biotechnology in India include among many others:

- Eradication of Rinderprest in India
- Indigenous technology for foot and mouth disease vaccine production
- Live attenuated vaccine for sheep and goat plague (PPR)
- Diagnosed avian influenza and controlled it by continuous surveillance
- Vaccine against avian influenza using indigenous strains
- Technical and scientific support to National FMD control programme
- Hundred percent import substitution for FMD diagnosis and control
- Diagnostic and Vaccine for equine virus infection.
- Diagnostic kit for GI parasites

The research work is carried out by scientists working in various Institutes under the umbrella of Indian Council of Agricultural Research. Some of the Institutes are listed as below.

- Indian Veterinary Research Institute, Izatnagar, Bareilly, U.P.
- National Dairy Research Institute, Karnal, Haryana
- National Bureau of Animal Genetic Resources, Karnal, Haryana
- Central Institute for Research on Buffaloes, Sirsa Road, Hisar, Haryana
- Central Institute for Research on Goats, Makhdoom, Farah, Mathura, U.P.
- National Institute of Animal Nutrition and Physiology, Bangalore, Karnataka
- Project Directorate on Cattle, Meerut, U.P.
- Project Directorate on Meat, Hyderabad
- Central Avian Research Institute, Izatnagar, Bareilly
- Project Directorate on Poultry, Hyderabad
- National Research Centre on Yak, Dirang, Arunachal Pradesh
- National Research Centre on Mithun
- National Research Centre on Pig.

- Central Sheep and Wool Research Institute, Avikanagar, Rajasthan
- National Research Centre on Camel, Bikaner, Rajasthan
- National Research Centre on Equines, Hisar, Haryana
- PD on Animal Disease Monitoring and Surveillance, Bangalore, Karnataka
- PD on Foot and Mouth Disease, IVRI Campus, Mukteshwar

In addition to these Biotechnology research, work is also being carried out in the Agricultural Universities notably being Haryana Agricultural University, Hisar, Punjab Agricultural University, Ludhiana, G.B. Pant University of Agriculture and Technology, Pantnagar, Veterinary Colleges at Akola, Parbhani, Faizabad, Mhow, Jabalpur, Pondicherry, Ranchi, Patna, Bhubaneshwar, Tirupati, Hyderabad, Bangalore, Kerala, Bikaner, Jammu, Veterinary and Animal Science Universities at Chennai, Mathura, and many others are carrying out teaching and research activities in the field of biotechnology in general and applications of it in agriculture and animal husbandry in particular. The State Agricultural Universities are major partners in the growth and development of agricultural research and education under the national agricultural research system. Currently, there are 41 State Agricultural Universities in India. These agricultural universities are responsible for research, training and dissemination of agriculture related information in the State. They generate new technologies to increase production, provide degree and certificate programmes in agriculture and help in the transfer of technology by participating in farmer training classes organised by local agricultural bodies.

Besides ICAR, State Agricultural Universities, Veterinary Colleges, some of the CSIR institutes and other important institutes are engaged in research on Biotechnology. Some of these are enumerated as follows.

- Indian Institute of Science, Bangalore
- Centre for Cellular and Molecular Biology, Hyderabad
- Central Drug Research Institue, Lucknow
- JALMA, Agra
- Defence Institute of Physiology and Allied Sciences
- Jawahar Lal Nehru University, New Delhi

- Delhi University
- Institute of Microbial Technology, Chandigarh
- Central Food Technological Research Institute, Mysore
- Institute of Genomics and Integrative Biology, Delhi
- Indian Institute of Chemical Biology
- Bose Institute, Kolkata
- Madurai Kamraj University, Madurai
- Poona University, Pune
- Indian Agricultural Research Institute, New Delh
- National Institute of Immunology, New Delhi

Biotechnology is a set of rapidly emerging and far-reaching new technologies with great promise in areas of sustainable food production, nutrition security, health care and environmental sustainability. The social impact of biotechnology is evident from India assuming a dominant place in vaccine exports, diagnostics, transgenics (BtCotton) and a number of biotherapeutics. There is a projection of an annual turnover of US $ 10 billion for India by 2010 and a speculated about 25% annual growth rate between 2010 and 2015 (India 2007). In the area of health care, new vaccines and diagnostics have been-indigenously developed and are under clinical trials. A major initiative has been taken to develop stem cell research in the country. Cutting edge research in areas of tissue engineering, bio-medical devices, biomaterials, nano-biotechnology and RNAi is begin supported. In the area of agriculture biotechnology the focus is on nutritional-enhancement, increased productivity and development of crops resistant to biotic and abiotic stresses. The Indian Government efforts are being focused to create circumstances for increasing access of common people to new technologies and products and promoting the mass use of these technologies for health care, nutritional security, employment generation and environmental well being. Life science and biotechnology sector is characterized by dynamic changes in flow of new idea and conception and development of new tools for research. Rapid responses are required to meet these challenges.

National Bioresource Development Board

Government of India has set up a National Bioresource development board with the programmes primarily focused in the area of biodiversity characterization and inventorization, bioprospecting, improvement and utilization of resources and capacity building.

Animal Biotechnology

Efforts continued for the improvement of animal productivity, development of newer animal vaccines and diagnostics, molecular characterization of indigenous breeds of livestock and "development of animal by products. New programmes have been initiated in the area of animal nutrition and development of newer animal vaccines. Standards was developed for estimation of mycotoxins in animal feed and distributed to various laboratories for routine analysis.

Biotechnology Parks and Incubators

Several biotech parks have been established in the country to promote the biotechnological activities in the country. The biotech parks and bitoech incubation centres provide an excellent template for the promotion of biotech start-up companies and the promotion of public private-partnership.

Besides these there are important institutions involved in biotechnology research activities in the country. Some of them are as follows.

National Institute of Immunology, New Delhi

This is a premier Institute carrying out pathbreaking researches in the frontline areas of biotechnology.

Bose Institute, Kolkata

Jawahar Lal Nehru University, New Delhi

National Centre For Cell Science, Pune

The Centre has emphasis on R&D activities in the areas of cell biology including stem cell biology, signal transduction, cancer biology, diabetes, and infection and immunity, chromatin architecture and gene regulation.

Centre for DNA Fingerprinting and Diagnostics (CDFD) Hyderabad

The Centre for DNA Fingerprinting and Diagnostics (CDFD) is an autonomous organization funded by the Department of biotechnology, Ministry of Science and Technology, Government of India. CDFD has been providing services for DNA fingerprinting, diagnostics, new born screening and bioinformatics based modern high technology DNA based methods, of direct benefit to the public, as well as in performing fundamental research of international standards in frontier areas of biological science. CDFD is providing DNA fingerprinting services to various Government and Law Enforcement Agencies and signed MoUs with State/Central-Forensic Science laboratories to popularize this technology for the benefit of the society. The major thrust areas of research in the centre are on infectious disease pathogens including *M. tuberculosis*, *H. pylori*, HIV, and HPV, silkmoth genetics and genomics, computational biology and bioinformatics; and fundamental studies on transcription and signal transduction.

National Institute for Plant Genome Research (NIPGR) New Delhi

Institute of Bioresources and Sustainable Development, Imphal

The research programmes of the institute have continued towards bioresource development and their sustainable use through biotechnological interventions for the socio-economic growth of the North East region.

Institute of Life Sciences, Bhubaneshwar

This institute is working on the molecular biology of the aging process, pathogenesis of chronic myeloid leukemia, infectious diseases such as cholera, malaria and filariasis and-in plant and environmental biotechnology.

Bharat Immunologicals and Biologicals Corporation Limited

There are two public sector undertakings i.e. Bharat Immunologicals and Biologicals Corporation Limited, (BIBCOL) and Indian Vaccines Corporation Limited (IVCOL). The BIBCOL located at Bulandshahar manufactures Oral Polio Vaccine being used in the National Immunization Programme. The IVCOL was established as a joint venture company. Efforts are being made to revive it with new products mix and financial pattern.

International Centre for Genetic Engineering and Biotechnolgoy (ICGEB), New Delhi

ICGEB is a premier Institute engaged in research efforts in identified areas of human health, agriculture and product development.

Animal Biotechnology Research Abroad

Almost every country in the world is having expectations from biotechnology research but still most of the research is costly and is carried out in developed countries of Americas, Oceania and Europe. In Asia, India, China and Japan are three major giants in biotechnology research. Some of the important research institutes in the field of animal biotechnology are as follows.

Animal and Natural Resources Institute, Beltsville, MD 20705 USA(http://www.ars.usda.gov/Main/site_main.htm). The Animal and Natural Resources Institute (ANRI) is located at the Henry A. Wallace Beltsville Agricultural Research Center (BARC) in Beltsville, Maryland. The research mission of the ANRI is to conduct research and to develop technology transfer programs that ensure high quality and safe food while protecting the natural resource base and the environment.

Centre for animal biotechnology, University of Melbourne, Australia. The Centre for Animal Biotechnology (CAB) was established in 1990 as a centre of excellence for large animal research within the School of Veterinary Science, The University of Melbourne. The major goal of CAB is to promote and conduct high quality research in large animal species that will contribute significantly to the development of improved methods for livestock management and disease control as well as serving as useful models for human disease processes.

The Institute for Animal Health (IAH), U.K. is a world-leading centre of excellence for research into infectious diseases of farm animals, sponsored by the Biotechnology and Biological Sciences Research Council.

The Friedrich-Loeffler-Institute, Federal Research Institute for Animal Health (FLI) is the national research centre for animal health of Germany. The institute was founded in 1910 and named for its founder Friedrich Loeffler in 1952. Among the animal diseases under research are for instance foot and mouth disease, influenza virus A, and avian influenza.

National Veterinary Institute, Technical University of Denmark, Bülowsvej 27, 1790 København V,Denmark (http://www.vet.dtu.dk/English.aspx). This Institute conduct research in infectious diseases in livestock and make diagnoses in diseased animals. It serves as a referred laboratory in this part of the world. Hybridoma laboratory of this institute generate and produce monoclonal antibodies for in house diagnostics or research.

Kimron Veterinary Institute, Israel (www.agri.huji.ac.il/~yakobson/vetserv/kvi.htm): The KVI was established as a small diagnostic laboratory during the British mandate in 1928. The KVI is the diagnostic and research arm of the Veterinary Services of the Ministry of Agriculture. The laboratories are allocated to 5 divisions (Division of Pathology, Division of Avian and Fish Diseases, Division of Parasitology, Division of Bacteriology, Division of Virology.

The University of Veterinary Medicine Vienna, Austria

Besides these, Korean National Livestock Research Institute, National Agricultural and Veterinary Biotechnology Centre of Ireland, Swiss Federal Institute of Technology, Veterinary Laboratories Agency of UK, CSIRO Livestock Industries are important centres of research in animal biotechnology. Most universities in the US and all over the world have come up with biotechnology departments, which are engaged in teaching or research on various aspects of biotechnology.

Most of the research is carried out by biotechnology companies besides government sponsored research Institutes and Universities. Approximately 105 companies have been identified to be involved in animal biotechnology in the US (http://www.bioportfolio.com). These are a mix of animal healthcare companies and biotechnology companies.

8

Techniques Used in a Biotechnology Laboratory

r DNA Biotechnology

The concept of modern biotechnology had its initiation with the advent of recombinant DNA which greatly increased the power of biotechnology and fuelled the imagination of scientists and laymen alike. Applications of biotechnology are flourishing particularly in the fields of human health and agriculture. Concomitantly, however, growing concerns about potential risks and ethics have been debated in scientific and public fora and have sometimes led to moratoria issued by the scientific community itself or imposed by legislative powers.

Concept of r DNA

The discovery of structure of DNA helix in the 1950's, the unraveling of genetic code in the 1960's, and the development and refinement of the tods of DNA engineering in the 1970's has revolutionized biological science. Using enzyme as genetic scissors' the genetic material can be snipped apart and reconstructed in combinations impossible to achieve by natural reproduction. Today, we can not only alter existing genes but construct totally different synthetic genes to make the organism to perform desired functions. The r DNA technique involves deliberately combining specific DNA sequences of one organism with DNA molecule of a different organism. The ability to recombine DNA from 2 different organisms is fundamental to genetic engineering. DNA segments of an organism (gene) can be isolated, sequenced, and synthesized. Reading the sequence of the letters in the instruction is called sequencing (Purchase, 1986). Sequencing and synthesis of genes are powerful tools in determining a gene function. Each gene contains triplet of base pairs called 'codons' that code for a specific amino acid and many codons are present in a gene code for production of a specific protein. Today we can identify the portions of a virus that are antigenic to which antibody it binds, and the viral genes that are responsible for pathogenicity. Some restriction enzymes that cut at specific sites leave single stranded protrusions called 'sticky ends'. Ends stick to one another because the bases are complementary. Two different DNAs cut by the same enzyme become 'annealed' by base pairing. These stands can be permanently bound by ligating enzymes called ligases. When the recombinant molecules replicate, there is corresponding amplification of the foreign DNA. This is called 'molecular cloning of genes' when the replicating recombinant molecules are inserted into an organism where the genetic information can be expressed, the protein specified by the foreign DNA is produced. This can be achieved by incorporating the foreign DNA into a plasmid which serves as a cloning vector for introducing the foreign DNA into bacterial or other cells. Hosts for protein production (vectors) can be bacteria, bacteriophage, yeasts, animal cells and even intact animals. The most widely publicized and most widely understood application of this technology was the production of foot and mouth disease vaccine (FMD). Since the virus is a RNA-Virus, complementary DNA (cDNA) was synthesized using viral RNA as a template before insertion into the vector. Oncogenic retroviruses, SV 40, Vaccinia, fowl pox, adeno and bacula (an insect virus) viruses have also been used as Vectors. When the vector virus multiplies in the host, it produce, not only its own protein, but also those of the other virus.

There are a number of techniques which are routinely used in a biotechnology laboratory. Basic to all biotechnology research is the ability to manipulate DNA. First and foremost for recombinant DNA work, researchers need a method to isolate DNA from different tissues or organsisms. To release the DNA from a cell, the cell membrane

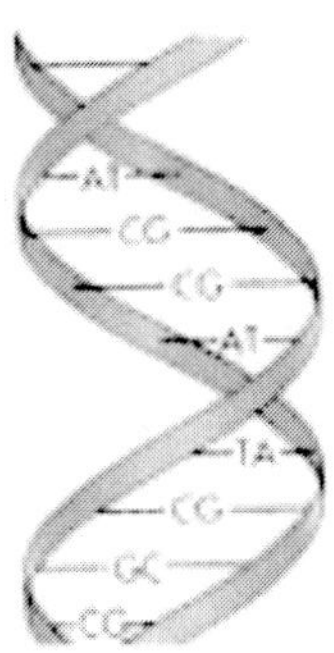

must be destroyed. Tissue samples from animals have to be ground up to release the intracellular components. Once released , the intracellular components are separated from the remains of the outer membranes by centrifugation or chemical extraction. Chemical extraction uses the properties of phenol to remove unwanted proteins from the DNA. Use of enzyme ribonuclease is made to separate RNA from DNA. Once separated electrophoresis separates DNA fragments by size. A simple method has been described below.

DNA Preparation from Tissue

1. Put 60 to 80 mg of tissue in a petri dish with culture media and divide the tissue into two pieces.
2. Put the tissue into two sterile 15 ml tubes and centrifuge for 2 min at 4°C at 1500 rpm.
3. Remove the supernatant, and wash twice with 1 ml 1X PBS or DNA-buffer.

   (It is possible to store the pellet at -80°C; in that case, add 1 ml 1X PBS and resuspend the pellet. Use a cryo-tube and centrifuge at 1500 rpm for 2 min at 4°C. Remove the supernatant, and freeze the pellet.)
4. Remove supernatant and resuspend the pellet in 2.06 ml DNA-buffer.
5. Add 100 μl proteinase K (10 mg/ml) and 240 μl 10% SDS, shake gently, and incubate overnight at 45°C in a waterbath.
6. If there are still some tissue pieces visible, add proteinase K again, shake gently, and incubate for another 5 hr at 45°C.
7. Add 2.4 ml of phenol, shake by hand for 5 to 10 min, and centrifuge at 3000 rpm for 5 min at 10°C.
8. Pipette the supernatant into a new tube; add 1.2 ml phenol, and 1.2 ml chloroform/isoamyl alcohol (24:1); shake by hand for 5 to 10 min, and centrifuge at 3000 rpm for 5 min at 10°C.
9. Pipette the supernatant into a new tube, add 2.4 ml chloroform/ isoamyl alcohol (24:1), shake by hand for 5 to 10 min, and centrifuge at 3000 rpm for 5 min at 10°C.

10. Pipette the supernatant into a new tube, add 25 μl 3 M sodium acetate (pH 5.2) and 5 ml ethanol, and shake gently until the DNA precipitates.
11. Take a glass pipette, heat it over a gas burner, and bend the end to a hook. Fish the DNA thread out of the solution using the hook and transfer DNA to a new tube.
12. Wash the DNA in 70% ethanol and dry it in the speed vac.
13. Dissolve the DNA in 0.5 to 1 ml sterile water overnight (or longer if necessary) at 4°C on a rotating shaker.
14. Measure the DNA concentration in a spectrophotometer and run 200 ng on a 1% agarose gel.

Reagents

- Chloroform
- EDTA, 0.5 M
- Ethanol, absolute
- Isoamyl alcohol
- Sigma, Cat. I-3643
- Phenol
- Phosphate Buffered Saline (PBS), 1X
- Proteinase K
- EM Science, Gibbstown, WV Cat. 24568-2 (100 mg)
- RNase A
- Boehringer Mannheim, Cat. 109 169
- Sodium dodecyl sulfate (SDS) solution, 10%
- Digene Diagnostics, Beltsville, MD, Cat. 3400 to 1016.

Preparation

DNA Buffer (Tris-EDTA)

1 M Tris pH 8.0 20 ml

0.5 M EDTA 20 ml

Sterile water 100 ml

Proteinase K (10mg/ml)

Dissolve 100 mg Proteinase K in 10 ml TE for 30 min at room temperature (RT), and store Aliquot at -20°C.

RNase A (20 mg/ml)

Dissolve 200 mg RNase A in 10 ml sterile water, boil for 15 min, and cool to RT and store Aliquot at –20°C.

Toe DNA Preparation for Southern Blots

Cut the last phalange of one toe into 0.5 ml of lysis buffer (100 mM Tris pH 8.0, 200 mM NaCl, 5 mM EDTA, 0.2% SDS, 200-500 μg/ml Proteinase K) and incubate at 50-55°C for 24-48 hours with gentle inversion of the tube.

Extract with 0.8 ml of phenol, microfuge 3 minutes. Transfer upper phase to a tube that contains 0.5 ml isopropanol. Cap the tube and invert 7 or 8 times. The DNA will form a wispy clot, that will either attach to an air bubble and float to the top, or will sink to the bottom.

Immediately fish out the clot on the outside of a yellow tip by swirling the tip in the clot. Expel any isopropanol that is in the tip, wait about 3 seconds to allow the remaining alcohol to evaporate, and then resuspend the DNA in 100 μl of TE in a fresh tube by pipetting up and down about 10 times.

Each toe produces enough DNA for 2 lanes on a Southern. Put 50 μl of DNA in a 200 μl volume overnight. EtOH ppt., resuspend and load on gel.

Southern Blot

Localization of particular sequences within genomic DNA is usually accomplished by the transfer techniques described by Southern (1975). Genomic DNA is digested with one or more restriction enzymes, and the resulting fragments are separated according to size by electrophoresis through an agarose gel. The DNA is then denatured *in situ* and transferred from the gel to a solid support (usually nitrocellulose filter or nylon membrane). The relative positions of the DNA fragments are preserved during their transfer to the filter. The DNA attached to the filter is hybridized to radiolabeled DNA or RNA, and autoradiography is used to locate the positions of bands complementary to the probe.

PCR

Polymerase Chain Reaction

The development of polymerase chain reaction-(PCR) has revolutionized the field of molecular biology. The technique consists basically of the enzymatic synthesis of millions of copies of a target DNA sequence (Saiki *et al.,* 1985). Using a thermostable DNA polymerase (Saiki *et al.,* 1988), and a succession of cycles that includes denaturation of the template DNA, hybridization of specific DNA primers to the template and extension of the primers, it is possible to generate multiple copies of the target region enzymatically. Thus, PCR provides a method for obtaining large quantities of specific DNA sequences from small amounts of DNA, including degraded DNA samples. PCR is an important tool for molecular biologists which is based on the *in vitro* enzymatic amplification of DNA sequences. Saiki *et al.* (1985) reported the PCR technique in the beginning. The polymerase chain reaction (PCR) is an in vitro enzymatic method which allows several million fold amplification of a specific DNA sequence. Since its introduction in 1985, PCR has facilitated the development of a variety of nucleic acid based detection systems for bacterial, viral and other pathogens as well as for genetic disorders. Owing to its high sensitivity, specificity and speed, PCR offers advantages over conventional diagnostic methods. A variety of PCR based assays have been described in the literature for the detection of infectious agents affecting man and animals.

Other recently developed DNA amplification methods, such as ligase chain reaction (LCR) ad 0 B replicate amplification have so far found only limited application but show promise for specific diagnostic applications. PCR already finds relatively broad use in the routine diagnosis of hereditary diseases of animals, such as bovine leucocyte adhesion deficiency (BLAD) and porcine malignant hyperthermia syndrome.

The purpose of a PCR (Polymerase Chain Reaction) is to make huge number of copies of a gene. This is necessary to have enough starting template for sequencing.

The Cycling Reactions

There are three major steps in a PCR, which are repeated for 30 or 40 cycles. This is done on an automated cycler, which can heat and cool the tubes with the reaction mixture in a very short time.

Denaturation at 94°C

During the denaturation, the double strand melts open to single stranded DNA, all enzymatic reactions stop (for example: the extension from a previous cycle).

Annealing at 54°C

The primers are jiggling around, caused by the Brownian motion. Ionic bonds are constantly formed and broken between the single stranded primer and the single stranded template. The more stable bonds last a little bit longer (primers that fit exactly) and on that little piece of double stranded DNA (template and primer), the polymerase can attach and starts copying the template. Once there are a few bases built in, the ionic bond is so strong between the template and the primer, that it does not break anymore.

Extension at 72°C

This is the ideal working temperature for the polymerase. The primers, where there are a few bases built in, already have a stronger ionic attraction to the template than the forces breaking these attractions. Primers that are on positions with no exact match get loose again (because of the higher temperature) and don't give an extension of the fragment.

Because both strands are copied during PCR, there is an exponential increase of the number of copies of the gene.

DNA Sequencing

The purpose of sequencing is to determine the order of the nucleotides of a gene. For sequencing, we don't start from gDNA (like in PCR) but mostly from PCR fragments or cloned genes.

The Sequencing Reaction

There are three major steps in a sequencing reaction (like in PCR), which are repeated for 30 or 40 cycles.

Denaturation at 94°C

During the denaturation, the double strand melts open to single stranded DNA, all enzymatic reactions stop (for example: the extension from a previous cycle).

Annealing at 50°C

In sequencing reactions, only one primer is used, so there is only one strand copied (in PCR : two primers are used, so two strands are copied).

Extension at 60°C

This is the ideal working temperature for the polymerase (normally it is 72°C), but because it has to incorporate ddNTP's which are chemically modified with a fluorescent label, the temperature is lowered so it has time to incorporate the 'strange' molecules. The primers, where there are a few bases built in, already have a stronger ionic attraction to the template than the forces breaking these attractions. Primers that are on positions with no exact match, come loose again and don't give an extension of the fragment.

The bases (complementary to the template) are coupled to the primer on the 3'side (adding dNTP's or ddNTP's from 5' to 3', reading from the template from 3' to 5' side, bases are added complementary to the template).

When a ddNTP is incorporated, the extension reaction stops because a ddNTP contains a H-atom on the 3rd carbon atom (dNTP's contain a OH-atom on that position). Since the ddNTP's are fluorescently labeled, it is possible to detect the color of the last base of this fragment on an automated sequencer.

Because only one primer is used, only one strand is copied during sequencing, there is a linear increase of the number of copies of one strand of the gene. Therefore, there has to be a large amount of copies of the gene in the starting mixture for sequencing. Suppose there are 1000 copies of the wanted gene before the cycling starts, after one cycle, there will be 2000 copies : the 1000 original templates and 1000 complementary strands with each one fluorescent label on the last base, after two cycles, there will be 2000 complementary strands, three cycles will result in 3000 complementary strands and so on.

Separation of the Molecules

After the sequencing reactions, the mixture of strands, all of different length and all ending on a fluorescently labelled ddNTP have to be separated; on an acrylamide gel, which is capable of separating a molecule of 30 bases from one of 31 bases, but also a molecule of 750 bases from one of 751 bases. All this is done with gel electrophoresis. DNA has a negative charge and migrates to the positive side. Smaller fragments migrate faster, so the DNA molecules are separated on their size.

Detection on An Automated Sequencer

The fluorescently labelled fragments that migrate through the gel, are passing a laser beam at the bottom of the gel. The laser exites the fluorescent molecule, which sends out light of a distinct color. That light is collected and focused by lenses into a spectrograph. Based on the wavelength, the spectrograph separates the light across a CCD camera (charge coupled device). Each base has its own color, so the sequencer can detect the order of the bases in the sequenced gene.

Assembling of The Sequenced Parts of a Gene

For publication purposes, each sequence of a gene has to be confirmed in both directions. To accomplish this, the gene has to be sequenced with forward and reverse primers. Since it is only possible to sequence a part of 750 till 800 bases in one run, a gene of, for example 1800 bases, has to be sequenced with internal primers. When all these fragments are sequenced, a computer program tries to fit the different parts together and assembles the total gene sequence.

Freezing of Buffalo Semen (Jindal, 1995)

Collection of SEMEN : The semen from buffalo bull is collected using a shorter sized artificial vagina as compared to cattle. Another male bull can be used as a teaser. The semen is examined for mass activity on warm stage (37°C) of a microscope and graded on 1 to 4 scale. The semen with a mass activity of +3 and above is selected for freezing. The average concentration of sperms in buffalo semen is 1200 million per ml. A frozen semen dose is expected to contain a total of 50 million sperms or at least 15 to 20 million live sperms. Only one sperm is needed for fertilization. The following dilution rats are used thus reducing the time spent to determine sperm concentration and also obviating the use of a costly spectrophotometer. However, if available, the determination of sperm concentrations by a spectrophotometer is advised and desirable.

Consistency	Overall Dilution Rate
Watery or Thin	10
Creamy	12
Thick Creamy	14

An average ejaculate of 2.5 to 3 ml so diluted semar becomes 25 to 40 ml and can be used to fill about 50 to 80 straws of 0.5 ml capacity. If two ejaculates are collected, about 100 to 160 straws can be frozen from a buffalo bull in a day. Twice weekly collections are advised. This schedule can optimally give you 1000 straws per month or 10,000 straws per year. Normally, seasonal variation and various other factors reduce the number considerably.

Tris, Citric Acid and Fructose diluter is used for dilution of buffalo semen. The dilution contains *Penicillin* and *Streptomycin* as antibiotics.

Egg yolk from fresh eggs is added at the level of 10%. This level is less than the level recommended for cattle (20%) and is quite suitable for buffaloes. It increases the visibility of sperms under the microscope. The buffer is autoclaved as also all the glasswares etc used in the freezing process.

The total volume of buffer needed for dilution is divided into two parts.

a) Non glycerolated
b) Glycerolated

Half of the buffer required for dilution is kept separately and is added with 12 % glycerol so that the glycerol level in the final diluted semen is 6%. The use of 6 per cent glycerol is beneficial as compared to 7% advocated in case of cattle (reduced backward movement in the post thaw semen and ensures better freezability). The part A and part B are kept under water in a cold handling cabinet and allowed to cool till 5°C.

This cooling should take about 1½ to 2½ hours. The glycerolated and non glycerolated portions are mixed at 5°C. The mixed and diluted semen is allowed to remain in water bath at 5°C for 4 hours to equilibrate. The diluted semen is filled in 0.5 ml straws in a cold handling cabinet taking care that there is minimum temperature variation.

If the semen is required to be filled in 0.25 ml mini straws, then accordingly the dilution rate is reduced to half. All other steps remain the same.

The straws can be filled using a vacuum pump and a bubbler comb or using an automatic filling and sealing machine. The automatic filling and sealing machine reduces friction and associated rise in temperature.

The semen is examined under the warm stage of microscope before filling in straws. Semen samples showing less than 70% motility are discarded and not filled.

The straws are exposed to liquid nitrogen vapours for 13 minutes in a simple termocole box. The approx internal size of the box is length 1.2 feet, width 8 inches and height 9 inches. Liquid nitrogen at a level of 2 mm is kept in the thermocole and straws are kept 3 mm above the liquid nitrogen levels. The straws are finally dipped in liquid nitrogen. The following day representative three to four straws are thawed in a water bath at 40°C for 30 sec. and evaluated for post thaw motility on a warm stage microscope. Semen samples with 30% or more post thaw motility are stored in biological storage cryocans and those with less motility are discarded.

The liquid nitrogen containers are periodically checked for liquid nitrogen levels and replenished with liquid nitrogen as and when needed.

The above method is recommended for use in the field as well as in a laboratory using minimum equipment and incurring less expenditure (Jindal, 1994).

Tissue Culture

The culture of whole organs, tissue fragements and cells in a suitable nutrient medium is called tissue culture. Tissue culture provides the basis of studying the cell function in carefully controlled conditions. Cell culture is removal of cells (eg oocyte, embryo) and placing them in artificial environment conductive to growth. The environment consists of a liquid or semi-solid medium that provides environment (eg anaerobic) and nutrients (eg glucose) and ions etc essential for survival and growth. There are a number of tissue culture medium which have been formulated keeping these points in view. The medium is usually fortified with special growth factors, chemicals etc for specific purposes and with antibiotics to prevent growth of harmful bacteria. The medium is usually isotonic and contains a buffering agent to prevent harmful shifts in pH. Tris or Phosphate buffers are most common buffering agents. *Penicillin, Streptomycin,* and/or *Heomycin* are the most common antibiotics added to tissue culture media. The water used for tissue culture media preparation is usually triple glass distilled or milli Q purified depending on the application and availability. Culture of undifferentiated cells like stem cell culture and culture of differentiated cells like embryo culture are some of the examples where culture is practiced. Vaccines like for polio, measles, mumps, chickenpox etc are made by cell culture. Cell culture is also used for large scale production of cells that have been genetically engineered to produce proteins having medicinal or commercial value.

For cell culture natural media like biological fluids have been used whereas artificial media are more preferred now a day.

Cell culture medium consists of a basic medium which satisfies cellular requirement for nutrients, salts and pH maintenance and suitable supplementary additions like growth factors, serum etc., which permit the growth of cells in a optimum way. Simulating natural conditions say for example oocyte or embryo culture requires the conditions present in the uterus i.e., anaerobic conditions, which can be met with the use of a carbon dioxide incubator and a tissue culture medium containing estrus goat serum or fetal calf serum, which provides some of the essential nutrients. Besides serum, other biological fluids have been used as media eg coconut milk as semen extender, amniotic fluid, aqueous humour, insect haemolymp etc.

Preparation of Media

Normal saline

For	1000 ml
Sodium chloride	8.5 gm
Streptomycin sulfate	100 mg
Penicillin G sodium	100000 IU
Distilled water	1000 ml

Dulbecco's phosphate buffered saline (D-PBS)

For 1000 ml	pH 7.2 - 7.4
Osmolality	270 - 290
Magnesiun chloride (Anhydrous)	100 mg
Potassium chloride	200 mg
Potassium phosphate monobasic (Anhydrous)	200 mg
Sodium chloride	8000 mg
Sodium phosphate dibasic (Anhydrous)	1150 mg
Calcium chloride dihydrate	100 mg
D-Glucose	1000 mg
Sodium pyruvate	36 mg
Streptomycin sulfate	50 mg
Penicillin G sodium	100000 IU
Phenol red	10 mg
Distilled water (added up to)	1000 ml

Oocyte collection medium

For 100 ml	pH 7.2 - 7.4
Osmolality	270 - 290
Dulbecco;s phosphate buffered saline	90 ml
Estrous goat serum (heat inactivated)	10 ml
Bovine serum albumin (Fraction V)	100 mg

Oocyte holding medium

For 100 ml	pH 7.2 - 7.4
Osmolality	270 - 290
TCM-199 (Hepes) (M 7528, Sigma)	90 ml
EGS (Heat inactivated)	10 ml
BSA (Fraction V)	3 mg/ml
L-Glutamine	100 μg/ml
Gentamicin	50 μg/ml
Sodium pyruvate	0.25 mM/ml

Oocyte maturation medium

For 5 ml	pH 7.2 - 7.4
Osmolality	270 - 290

TCM-199 (NaHCO3) (M 4655, Sigma)	4 ml
EGS (Heat inactivated)	1 ml (20%)
L-Glutamine	100 μg/ml
Gentamicin	50 μg/ml
Sodium pyruvate	0.25 mM/ml

Embryo development medium

For 10 ml	pH 7.2 - 7.4
Osmolality	270 - 290
TCM-199 (NaHCO3) (M 4655, Sigma)	9 ml
FBS	1ml
L-Glutamine	100 μg/ml
Gentamicin	50 μg/ml
Sodium pyruvate	0.25 mM/ml

Modified TALP

For 1000 ml	
NaCl	6.662 g
KCl	0.238 g
$NaHCO_3$	2.104 g
$NaH_2\ PO_4$	0.0408 g
$CaCl_2.2H_2O$	0.300 g
$MgCl_2.6H_2O$	0.100 g
Sodium lactate (60%)	1860 μl
Sodium pyruvate	27.5 mg
Streptomycin sulphate	50 mg
Penecilline-G-sodium	1,00,000 IU
Phenol Red	10 mg
Distilled water (add up to)	1000 ml.

Sperm TALP / Fert TALP

For 20ml	
TALP solution	16 ml
EGS (Heat inactivated)	4 ml (20%)
Heparin (H 1027, Sigma)	10 μg/ml

Laboratory Procedure for IVF (Kharche *et al.*, 2008)

Source of Oocytes

The oocytes for the IVM can be obtained either from ovaries of live animals or from ovaries collected from slaughter house depending on the application of IVF technology. Several methods like dissection and aspiration of follicles have been used for collection of oocytes. Depending upon the species, 18 to 21 gauge needles have been used by different workers (Deb and Goswami, 1990). The number of oocytes obtained from cow ovary averaged 16 out of which 80 % were suitable for culture.

Ultrasound guided or laparoscopic oocyte retrival is of major importance in the production and early propogation of transgenic for goats with lowered fertility. Whereas, oocyte retrieval form slaughter house ovaries is of choice for large scale economical production of early and late stages of embryos to be used in the various biotechnological experiments.

Preparation of Estrous Goat Serum

Goats are observed for the occurrence of estrus daily twice at 12 hour intervals in the experiment herd. Estrous serum from these goats is collected 12 h following the onset of estrus and filtered through millipore filters (with a pore size of 0.22 μm). The estrous goat serum (EGs) was heat inactivated at 56°C for 30 minutes in a water bath, dispensed into 1 ml and 10 ml aliquots and stored at -20°C until used.

Collection of Oocytes

The application of *in vitro* technology for maturation, fertilization and culture of domestic animal oocytes partially depends on reliable, repeatable and efficient technique for their recovery. In most cases the source of oocytes was ovaries derived from slaughter materials. Laparotomy and laparoscopy through a mid ventral incision of animals have also been used as an alternative to collect oocyte from living animals. The laparoscopic technique was very reliable and allowed for the recovery of predictable number of oocytes and also less invasive than standard surgery (laparotomy). It allowed direct view of the ovary for aspiration of superficial follicles with a reduced risk of injury to the ovary. A safe practical alternative to surgical or laparoscopic oocyte collection is ultrasound guided transvaginal aspiration technique, however in goats only a few references are available. It has a disadvantage that it did not allow visualization of all follicles on the monitor with reduced oocyte recovery and also increased the chances of ovarian injury.

Laparoscopic Ovum Pick-up

Donor goats are restrained on a standard laparoscopy table under general anaesthesia and follicles are aspirated under laparoscopic observation using a 20 gauge needle mounted in plastic pipette connected to a collection tube and vacuum line. To recover higher number of oocytes per LOPU session, the donor goats are stimulated with gonadotrophins before retrieval of the oocytes.

Oocyte Retrieval from Ovaries Collected form Slaughter House

Goat ovaries are collected from a local abattoir and transported within 3 hours to the laboratory in a warm saline solution (30-35°C), supplemented with 100 IU penicillin G and 100 µg streptomycin sulfate/ml for the recovery of oocytes. The surrounding tissues are trimmed off with fine scissor and the ovaries are washed with normal saline. Oocytes are recovered from these ovaries using following techniques.

Slicing Technique

The individual ovary is slice into small pieces with a scalpel blade in a sterilized petri dish containing oocyte collection medium (OCM). The oocytes are then pick-up and transfer in a petri dish containing oocyte holding medium (OHM).

Follicle Puncture Technique

In this technique oocytes are recovered by puncturing of 1 to 4 mm diameter surface follicles using a 18 Gauge needle, attached to a disposable syringe in a sterilized petri dish containing OCM.

Follicle Aspiration Technique

Oocytes are collected by aspiration of the visible surface follicles using 22 to 24 Guage needle, attached to a disposable syringe containing OCM. The aspirated oocytes are pour in to a sterilize petridish.

Follicle Isolation Technique

In this technique the entire follicle dissected out individually from the ovary in a petridish containing OCM under the stereo zoom microscope using sharp pointed iris scissors and fine pointed forceps. The isolated follicles were punctured with 18 Gauge needle under stereo zoom microscope to yield oocytes.

Evaluation of Oocytes

Recovered oocytes to be graded as excellent (A), good (B), fair (C) and poor (D) quality depending upon their cumulus investment and cytoplasmic distribution for *in vitro* maturation.

Excellent (A): Oocytes with more than four layers or bunch of compact cumulus cell mass with evenly granulated cytoplasm.

Good (B): Oocytes with at least two to four layer of compact cumulus cell mass with evenly granulated cytoplasm.

Fair (C): Oocytes with at least one complete compact layer of cumulus cell with evenly granulated cytoplasm.

Poor (D): Oocytes with no cumulus cell or incomplete layer of cumulus cell or expanded cumulus mass with or without dark or unevenly granulated cytoplasm.

Only oocytes of grades A and B are used further for *in vitro* maturation.

In vitro Maturation of Oocytes

The COC's of A and B grades are selected for *in vitro* maturation. These oocytes are washed in a 35 mm petri dish containing OHM followed by 10 times drop washing in the OHM. Ten to 15 cumulus intact COC's are transferred into a 50 µl drop of maturation media supplemented with 20% estrous goat serum (EGS), under warm mineral oil in a tissue culture dish (35 mm x 10 mm), equilibrated for 2h in a CO_2 incubator before the oocytes are added. These oocytes are cultured for 24 to 27h at 38.5°C under humidified atmosphere (90 to 95% relative humidity) of 5% CO_2 in air.

Preparation of Cells for Co-culture

Granulosa Cell Monolayer: Granulosa cell monolayers are prepare from granulose cells aspirated from small follicles, judge to be non-atretic. Masses of granulose cells are isolated following recovery of individual COC from follicular aspirates and are then washed through 4 to 6 drops (100 µl) of OHM. Cell suspensions (approximately $5x10^6$ cells/ml in 5µl) are culture in 100 µl drops of OHM under warm mineral oil for 48h at 38.5°C under humidified atmosphere (90 to 95% relative humidity) of 5% CO_2 in air. After 48h the expanded cumulus masses, now forming monolayers, are used for co-culture of presumptive zygotes and early 2-cell stage embryos.

Oviduct Epithelial Cell Monolayer : Caprine oviduct epithelial cell (cOEC) cultures are prepared simultaneously when the oocytes are harvested. Oviducts (4 to 6) of the ovaries having newly formed corpora lutea are obtained at abattoir, washed and wrapped in a piece of aluminum foil. The oviducts are brought in to the laboratory at 4°C in the ice box. The oviducts are cleaned of the excess fat and surrounded tissue and hold in thermos containing OCM with gentamicin sulfate (10mg/ml) at 25 to 30°C for 1 to 3h before use. The effluents of oviduct epithelial cells are collected either by squeezing out the content with sterilized glass slide/forceps or by repeated flushing with 10ml of OCM. The OEC are disaggregated in to small clusters by repeated flushing through a 26G needle and the resultant cell suspension are taken in to 15 ml centrifuge tubes. Cells are washed twice with OCM by centrifugation (500rpm/26 x g) for 5 minutes. The cell are layered in 100μl drops of OHM. After 24h, the cells clusters are washed four times in OHM and then transferred to the culture droplets (approximately 50 cell cluster/50 μl droplet) to be used subsequently for co-culture with cleaved embryos obtained from IVM and IVF.

Assessment of Nuclear Maturation After *in vitro* Maturation of Oocytes

At the end of the 24 hours incubation period the cumulus cells are removed completely from the oocytes by pipetting the complexes into and out a finally drawn pasture pipette with an inner diameter similar to that of a nude oocyte. Care is exercised to avoid rupturing the zona pellucida. The nude oocytes are incubated in hypotonic 0.075M KCl solution for 15 min at room temperature. The treatment results in swelling of the oocytes. Oocytes are then transferred on to clean, grease free slides onto which the ice cold fixative (acetic acid, methanol, 1:3) is added drop by drop. The fixative is added till the zona pellucida breaks open. The slides are then dried and stained for 3 to 5 min with 2% Giemsa solution, and subsequently rinsed with distilled water. The nuclear maturation/chromosomal configuration is assessed under a microscope at 1000 X magnification.

Analysis of Nuclear Status is Ascribed to One of the Following

Germinal vesicles (GV) containing a single large nucleus with uniformly distributed filamentous chromatin subsequently condensing to form a ring of condensed chromatin around the compact nucleous.

Germinal vesicle breakdown (GVBD) identified by disappearance of compact nucleoulus and nuclear membrane, and gradual condensation of chromatin material beginning to be distinguishable into chromosomal identities.

Metaphase-I (M-I) indicated by the formation of individual bivalents with chiasmata, subsequently terminalization of chiasmata, appearance of tetrads, separation of two chromosome sets towards opposite poles (anaphase-I) separation of two chromosome sets is completed (telophase-I)

Metaphase-II (M-II) denoted by emission of first polar body and with haploid set of chromosomes.

Sperm Preparation for *in vitro* Fertilization of Oocytes

Fresh semen is collected by an artificial vagina from a fertile buck. The ejaculates are visually examined for volume, color, consistency and mass motility. The semen samples having +5 motility are selected for the IVF. After assessing the motility, 50 µl of semen is taken in to a centrifuge containing 5 ml of sperm TALP medium and centrifuge at 1800 rpm for 5 to 10 minutes. The supernatant is discarded; the sperm pellet is resuspended with 5 ml of fresh TALP medium and centrifuge as before. Fifty micro liter of resulting pellet is then transferred into a vial having 750 µl of fertilization TALP supplemented with 20%EGS and 10µg/ml heparin and is kept in a CO_2 incubator for 30 to 45 minutes.

In vitro Fertilization of Oocytes

After 27h of *in vitro* maturation, cumulus cells are removed by treating with OHM containing 0.1% hyaluronidase and repeatedly passing them through a fine pipette. The oocytes are washed 5 to 10 times in fertilization TALP. After washing 15 to 20 oocytes are placed in each 50µl drop of fertilization TALP covered with warm mineral oil in tissue culture petri dish. After heparin treatment, sperm concentration is assessed in a hemocytometer/sperm meter. An aliquot of (25 to 50 µl) the sperm suspension is added to each fertilization drop to give a final concentration of 2-5x10^6 sperm/ml, oocytes are co-incubated with capacited spermatozoa for 18h at 38.5°C under humidified atmosphere (90-95% relative humidity) of 5% CO_2 in air.

Assessment of *in vitro* Fertilization of Oocytes

The oocytes are washed with TCM-199 +5% FBS after 18 to 20 h post insemination.

The oocytes are then transferred on a grease free glass slide and the extra medium removed with the help of filter paper.

A small drop of mountant (Vaseline was 1:1) is placed on the edges or corners of the cover slip.

The coverslip is placed over the oocytes.

The slide is then place in acetic acid:ethanol (1:3) in a slide holder for 24 hours.

The slide is removed after 24 hours of fixation and stained with 0.1% orcein.

The oocytes are examined under interference contrast or phase contrast microscope. The following classification is used for assessment of *in vitro* fertilization.

i) Only metaphase II plate: unfertilized.
ii) Only one pronucleus: fertilized, monospermy.
iii) Two pronuclei: fertilized, normal.
iv) More than two pronuclei: fertilized, polyspermy.

Assessment of Cleavage

The oocytes are evaluated under a zoom stereo microscope or phase contrast inverted microscope after 42 to 48 hours post insemination for evidence of cleavage. The stage of cleavage is judged by counting the number of blastomeres. The results are recorded in terms of cleavage rate i.e. percentage of embryos that have cleaved to the 2 cell stage or beyond.

In vitro embryo culture

After 18h of sperm-oocyte co-incubation, the oocytes are washed by repeated pipetting in embryo development medium (EDM) to remove adhering sperm cells. Inseminated ova/embryos are transferred on to granulose cell/OEC monolayer and cultured at 38.5°C under humidified atmosphere (90-95% relative humidity) of 5% CO_2 in air. After 48h post insemination, the cleavage rate is evaluated under a inverted phase contrast microscope. These embryos are further cultured until day after insemination.

Every 48h after insemination the EDM is refreshed by extracting 50 μl of old medium from the micro drops and adding 50 μl of fresh medium. From day 6 after insemination, the embryos are monitored daily to assess blasocyst development. At day 10 after insemination, the embryos are fixed to evaluate their embryonic stage.

References

Aitken, J., Fisher, H. (1994).Reactive oxygen species generation and human spermatozoa:the balance of benefit and risk.BioEssays 16:259-267

Aitken, R.J., Harkiss, D., Knox, W., Paterson, M., Irvine, D.S. (1998). A novel Signal cascade in capacitating human spermatozoa characterized by a redox regulated, cAMP-mediated induction of tyrosine phosphorylation. J.Cell. Sci. 111, 645-656.

Allen, W.R. (1982) Enbryo transfer in the horse. In : Mammalian Egg Transfer, Ed. C.E.Adams, CRC Press, Florida pp 135

Almquist, J. O., P. J. Glantz, and H. E. Shaffer. (1949). The effect of a combination of penicillin and streptomycin upon the livability and bacterial content of bovine semen. J. Dairy Sci. 32:183–190.

Alok, Deoraj, N.S. Haque & J.S. Khillon (1995). Transfer of human growth hormone gene into Indian Major Corp. (Labeo rohita) Int. J. Anim.Sci. 10 : 5-8.

Alvarez, E., Storey, BT. (1984).Lipid peroxidation and the reactions of superoxide and hydrogen peroxide in mouse spermatozoa. Biol.Reprod 30:833-841.

Alvarez, E., Storey, BT. (1992). Evidence for increased lipid peroxidative damage and loss of superoxide dismutase activity as a mode of sublethal cryodamage to human sperm during cryopreservation.J. Androl.13:232-241.

Amann, R. P., Seidel, Jr. G.E., and Brink, Z.A. (1999). Exposure of thawed frozen bull sperm to a synthetic peptide before artificial insemination increases fertility. *J. Androl.* 20:42–46.

Aminov, R.I., K. Kaneichi, T. Miyagi, K. Sakka and K. Gilbert, H.J., G.P. Hazlewood, J.I. Laurie, C.G. Orpin and Ohmiya, 1994. Construction of genetically marked G.P. Xue, 1992. Homologous catalytic domains in a Ruminococcus albus strains and conjugal transfer rumen fungal xylanase-Evidence for gene of plasmid pAM$1 into them. J. Ferm. Bioeng., 78: 1- duplication and prokaryotic origin. Mol. Microbiol., 6: 5.

Amoah, E. A., and S. Gelaye. (1997). Biotechnological advances in goat reproduction. J. Anim. Sci. 75:578–585.

Anand, S.R., Atreja, S.K., Chauhan, M.S., and Behl, R. (1989) In vitro capacitation of goat spermatozoa. Ind. J. Exp. Biol. 27(11)921-924

Andrew, J.C. and Bavister, B.D. (1988) Hamster zonae pellucidae cannot induce physiological acrosome reactions in chemically capacitated hamster spermatozoa in the absence of albumin. Biol. Reprod. 41: 117-122.

Anwar, M., Riaz, A., Ullah, N and Rafiq, M. (2008) Use of ultrasonography for pregnancy diagnosis in Balkhi sheep. Pakistan Vet. J. 28(3) 144-146

Aonuma, S., Okabe, M, Kishi, Y , Kawaguchi, M. And Yamada, H. (1982) Capacitation inducing activity of serum albumin in fertilization of mouse oocyte in vitro. J. Pharmacodynamics. 5:980-987

Austin, C.R. (1951) Observation on the penetration of sperm in the mammalian egg. Australian J. Sci. Res. 84: 581-589.

Baldassarre, H. and Karatzas, C.N. (2004) Advanced assisted reproduction technologies (ART) in goats. Animal Reproduction Science 82-84: 255-266.

Bandopadhyay, P.K. and Temin, H. (1984) Expression of complete chicken thymidine kinase gene inserted into a retrovirus vector. Mol. Cell. Biol. 4:749-754.

Bank, H. and Maurer, R.R. (1974) Survival of frozen rabbit embryos. Exp. Cell. Res. 89:188-196.

Bauman, D. E. (1999). Bovine somatotropin and lactation: From basic science to commercial application. Dom Anim Endo 17:101–116

Bedford, J.M. (1983) Significance of the need for sperm capacitation before fertilization in eutharian mammals. Biol Reprod. 28: 108-120.

Benchaib M, Braun V, Lornage J, Hadj S, Salle B, Lejeune H, et al. Sperm DNA fragmentation decreases the pregnancy rate in an assisted reproductive technique. Hum Reprod 2003;18:1023–28.

Bhattacharyya, P. and Srivastava, P.N. (1955) Studies on deep freezing of buffalo semen. Proc. Indian Sci. Cong. Part III, Abstract

Bilton, R.J. and Moore, N.W.(1976) In vitro culture storage and transfer of goat embryos. Aust. J. Biol. Sci. 29:125-129.

Bishop, S.C. and Morris, S.A. (2007) Genetics of disease resitance in sheep and goats. Small Ruminant Research 70: 48-59.

Bittle, J.L., Houghten, R.A., Alexander, H., Shinnick, M., Sutcliffe, I.G., Lerner, R.A., Rowlands, D.J. and Brown, F.(1982) Protection against Foot and mouth disease by immunization with a chemically synthesized peptide predicted from the viral nucleotide sequence. Nature 298: 30-33

Boman, P., Tieman, M.Van Zaane. D., Bonsma, A.A. anode Boes, G.F. (1990). Vet. Immunol. Immunopathol. 24:211-226.

Bondioli, K.R. and Wright, R.W.Jr. (1983) In vitro fertilization of bovine oocytes by spermatozoa capacitated in vitro. J Anim. Sci. 57: 1001-1005.

Bowman, J.G.P. and B.F.Sowell,(2003) Technology to complement forage based beef production systems in the west. J. Anim. Sci.81:E18-E26

Brackett, B. G., Bousquet, D., Nice, M. L., Donawick, W. J., Evans, J. F. and Dressel, M. A.: (1982)Normal development following in vitro fertilization in the cow. Biol. Reprod. 27:147-158,

Brackett, B. G., W. Boranska, W. Sawicki, and H. Koprowski. (1971). Uptake of hetrolougous genome by mammalian spermatozoa and its transfer to ova through fertilization. Proc. Natl. Acad. Sci. 68:353–357.

Bungum M, Humaidan P, Spano M, Jepson K, Bungum L, Giwerchman A. (2004) The predictive value of sperm chromatin structure assay (SCSA) parameters for the outcome of intrauterine insemination, IVF and ICSI. Hum Reprod;19:1401– 8.

Byrd, W. (1981) In vitro capacitation and chemically induced acrosome reaction in bovine spermatozoa. J. Expl. Zool. 215: 35-46.

Campbell, K.H.S; Mcwhir, J; ,Ritchie, W. and Wilmut, I. (1996). Sheep cloned by nuclear transfer from a cultured cell line Nature. 380: 64-66.

Cassou, R. 1964. La me´thode des pailletes en plastique adapte´e a' la ge´ne´ralisation de la conqe´lation. In: Proc. 5th Int. Congr. Anim. Reprod. Trento, Italy. 4:540–546 (Cited by Foote, 2002)

Cavanagh, D. (1986) Monoclonal antibody and nucleic acid probes for diagnosis of avian diseases. Av. Dis. 30:12-18.

Chandler, J.E., Painter, CL., Adkinson, R.W., Memon, M.A., Hoyt, P.G., (1988). Semen quality characteristics of dairy goats. J. Dairy Sci. 71, 1638-1646.

Chang, M. C. (1968). *In vitro* fertilization of mammalian eggs. J.anim . Sci. 27 (Suppl. 1):15-22.

Chang, M.C. (1959) Fertilization of rabbit ova in vitro. Nature (London) 184: 466-467.

Chantraprateep, P., Kobayashi, G., Lohachit, C., Virabil, P., Kunawongkrit, A., Techakum, Phu, M., Prateep, P. and Dustin, N. (1989) Success in embryo transfer in Thai Swamp buffalo Buffalo Bulletin 8(1) 4-5.

Chapaco, W., Ashton, N.W., Martel, R.K., Antonishyn, N. and Crosby, W.L. (1992). A feasibility study of use of random amplified polymorphic DNA in the population genetic and systematic of grass hoppers Genome. 35, 569-575.

Chikamatsu, N., Urakawa, M. , Fukui, Y., Aoyagi, Y. And Ono, M. (1989) In vitro fertilization and early development of bovine follicular oocytes matured in different culture system and treated with spermatozoa treated by different methods. Japanese J. Anim. Reprod. 35: 154-158.

Cocconcelli, P.S., E.Ferrari, F.Rossi and V.Bottazzi, (1992) Plasmid transformation of Ruminococcus albus by means of high voltage electroporation. FEMS Microbiol. Lett. 94: 303-208.

Cowie, A.T. (1969) General hormonal factors involved in lactogenesis. In Lactogenesis: the Initiation of milk secretion at Parturition, ed M.Reynolds and S.J.Folley, 157, Philadelphia, Univ. Penn. Press.

Crittenden, L.B. and Salter, D.W. (1986) Gene insertion : Current progress and long term goals. Av. Dis. 30:43-46.

Cruz, L.E., et al (1991) Successful transfer of Murrah buffalo embryo into Philippine Swamp buffalo recipients. Third World Buffalo Congress, Bulgaria Vol 3 : 586-590.

Dang, A.K. and Ludri, R.S. (2001) A study on the scanning electron microscopy of buffalo mammary gland. Asian Aust. J. Anim. Sci. 14(1) 13-15.

Davis, I. S., Bratton, R. W. and Foote, R. H. (1963). Livability of bovine spermatozoa at 5, -25, and -85°C in tris-buffered and citratebuffered yolk-glycerol extenders. *J. Dairy Sci.* 46:333–336.

Davis, K.J. (1987). Protein damage and degradation by oxygen radicals I: General aspects.J. Biol. Chem.262, 9895-9901.

Davis, R.O., Gravance, C.G., Overstreet, J.W., (1995). A standardized test for the visual analysis of human sperm morphology. Fertil. Steril. 63, 1058-1063.

Deb, S.M. and Goswami, S.L. (1990) Cytogenetic studies of goat oocytes cultured in vitro. International J. Anim. Sci. 5: 137-143.

Dodds, J.W. and Seidel, G.E. (1984) Effect of caffeine, Ca++ , capacitation time and strain on IVF in mice. Gamete Res. 10: 353-360.

Donoghue, A M, L A Johnston, U S Seal, D L Armstrong, R L Tilson, P Wolf, K Petrini, L G Simmons, T Gross and D E Wildt (1990) In vitro fertilization and embryo development in vitro and in vivo in the tiger (Panthera tigris). Biology of Reproduction 43: 733-744

Dresser, B.L. and Gelwickas, E.J., wachs, K.B. and K.L. (1988) Embryo cryopreservation and transfer in domestic cat. In. 11th International Congress on Animal Reproduction and Artifical Insemination, University College, Dublin, Ireland No 160, 3 pp.

Drost, M., Wright, J.M. Jr., Cripe, W.S. & Richter, A.R. (1983). Embryo transfer in water buffalo (Bubalus bubalis). Theriogenology., 20:579–584.

Durand, F.C. and G. Fonty, (2001). Establishment of W.J.M. Smith, G.P. Xue and C.S. McSweeney, (2001). cellulolytic bacteria and development of fermentative activities in the rumen of gnotobiotically-reared lambs receiving the microbial additive Saccharomyces cerevisiae CNM I-1077. Reprod. Nut. Dev., 41: 57-68.

Edwards JL, Schrick FN, McCracken MD, van Amstel SR, Hopkins FM, Welborn MG, Davies CJ. (2003)Cloning adult farm animals: a review of the possibilites and problems associated with somatic cell nuclear transfer. AJRI; 50:113–123

El-Manoufy, A.A., Seidel, A.A., Fattouh, El-SM and Abou Ahmed, M.M. (1986) Effect of caffeine on metabolic activity of epididymal spermatozoa of buffalo (Bubalus bubalis) Zuchthygiene. 21: 214-219.

Eng, L.A. , Kornegay, E.T. , Huntington, J. And Wellman, T. (1986) Effects of incubation temperature and bicarbonate on maturation of pig oocytes in vitro. J. Reprod. Fertil. 76: 657-662.

Eppig, J.J. and Schroeder, A.C. (1986) Culture systems for mammalian oocyte development: Progress and prospects. Theriogenology. 24: 97-106.

Erenpreiss J, Spano M, Erenpreisa J, Bungum M, Giwercman A. (2006) Sperm chromatin structure and male fertility: biological and clinical aspects. Asian J Androl;8:11–29.81:1289 –95.

Etherton, T. D. and D. E. Bauman. (1998). The biology of somatotropin in growth and lactation of domestic animals. Physiol Rev 78:745–761.

Evenson DP, Jost LK, Baer RK, Turner TW, Schrader SM. Individuality of DNA denaturation patterns in human sperm as measured by the sperm chromatin structure assay. Reprod Toxicol 1991;5:115–25.

Filatov MV, Semenova EV, Vorob'eva OA, Leont'eva O, DrobchenkoEA.(1999) Relationship between abnormal sperm chromatin packing and IVFresults. Mol Hum Reprod 5:825–30.

Flint, H.J. and K.P. Scott, (2000). Genetics of rumen microorganisms: gene transfer, genetic analysis and strain manipulation. In: Cronje, P.B. (Ed.), Ruminant physiology, CABI Publishing, Oxon, pp: 389-408. UK.

Foote, R. H. (1981). The artificial insemination industry. In: B. G. Brackett, G. E. Seidel, Jr., and S. M. Seidel (ed.) New Technologies in Animal Breeding. pp 13–39. Academic Press, New York.

Foote, R.H. and Bratton, R.W. (1949). The fertility of bovine semen cooled with and without the addition of citrate-sulfanilamideyolk extender. *J. Dairy Sci.* 32:856–861.

Foote, R.H. and Bratton, R. W. (1950). The fertility of bovine semen in extenders containing sulfanilamide, penicillin, streptomycin, and polymyxin. *J. Dairy Sci.* 33:544–547

Foote, R.H. (2002). The history of artificial insemination: Selected notes and notables. J Anim Sci. 80:1-10.

Foote, R.H., Parks, J.E. (1993). Factors affecting preservation and fertility of bull sperm: A brief review. Reprod. Fertil. Dev. 5, 665-673.

Fukui, Y. And Ono, H. (1989) Effects of sera, hormones and granulosa cells, added to culture medium in vitro maturation, fertilization , clevage and development of bovine oocytes. J. Reprod. Fertil. 86: 501-506.

Gilbert, H.J., G.P.Hazlewood, J.I.Laurie, C.G.Orpin and G.P.Xue (1992) Homologous catalytic domains in a rumen fungal xylanase- Evidence for gene duplication and prokaryotic origin Mol.Microbiol. 6:2065-2072.

Go, K.J. and Wolf, D.P. (1985) Albumin mediated changes in sperm sterol content during capacitation. Biol. Reprod. 32: 145-153

Goel, A.K. and Agrawal, K.P. (1992) A review of pregnancy diagnosis techniques in sheep and goats Small Ruminant Research 9, 255 - 264

Goel, A.K. Kharche, S.D and Jindal, S. K. (2009) Determination of early pregnancy and embryonic development by trans rectal ultrasonography in goats. Indian Journal of Animal Sciences 79:476-478

Goel,A.K. and Kharche, S.D. (2009) Advances in reproductive technique in goats- A review. Indian Journal of Small Ruminants 15(1)1-34.

Gordon, J. W., G. A. Scangos, D. J. Plotkin, J. A. Barbosa, and F. H. Ruddle. (1980). Genetic transformation of mouse embryos by microinjection of purified DNA. Proc. Natl. Acad. Sci. 77:7380–7384.

Guraya, S.S. (1977) Resumption of meiosis in mammalian oocytes cultured in vitro. Int. Rev. Cytol. 51: 49.

Hallerman, E.M. (1989). Genetic improvement of livestock through uti-lization of DNA level markers. Int. J. Animo Sci. 4 : 60-70.

Hein, W.R. and Harrison, G.B.L. (2005) Vaccines against veterinary helminthes. Vet. Parasitology 132: 217-222.

Hespell, R.D., D.E.Akin and B.A.Dehority (1997) Bacteria, fungi and protozoa of the rumen. In. Mackie, R.I., White, B.A. and Isaacson, R.E. (Eds) Gastrointestinal microbiology, Chapman and Hall, New York pp 59-141. U.S.A.

Ho,Y.W. and D.J.S. Barr (1995) Classification of anaerobic gut fungi from herbivores with emphasis on rumen fungi from Malaysia. Mycologia 87:655-677

Ijaz, A. And Hunter, A.G. (1989) Evaluation of Ca free Tyrodes capacitation medium for use in bovine in vitro fertilization. J. Dairy Sci. 72: 3280-3285.

Iritanai, A., Kasai, M, Niwa, K. And Song, H.B. (1984) Fertilization in vitro of cattle follicular oocytes with ejaculated spermatozoa capacitated in chemically defined medium. J. Reprod. Fertil. 70: 487-492.

Jain, G.C., Kumar, P., Khanna, S., Elsden, R.P. and Lohan, I.S. (1990) Calves born from embryo transfer technology in buffaloes. Int. J. Anim. Sci. 5:203-208.

Jasko, D.J., Lein. D.H., Foote, R.H., (1990). Determination of the relationship between sperm morphologic classification and fertility in stallion: 66 cases (1987-1988). J. Am. Vet. Med. Assoc. 197, 389-394.

Jin, D.I., Petters, R.M. and Im, K.S. (1994) Transgenic livestock-review. Asian Australasian Journal of Animal Sciences. 7:1-17.

Jindal, S. K.; Sajjan Singh; Pawan Singh and G.C. Jain (1994). A decade of embryo transfer technology in buffaloes. Problems and prospects. Indian Dairyman. 46' : 373-375.

Jindal, S.K. (1984) Goat Production, Cosmo Publications, New Delhi

Jindal, S.K. (1989). Relationship between some circulating hormones, metabolities and body composition in lactating crossbred cows and buffaloes. Ph.D. Thesis, NDRI, Karnal (Guide Dr.R.S.Ludri).

Jindal, S.K. (1995). A simple and practical method for the freezing of buffalo semen. Indian Dairyman 46(10) 767-768.

Jindal, S.K. and Ludri, R.S. (1990 a). Circulating thyroxine and triiodothyronine levels in lactating crossbred cows and buffaloes as affected by stage of lactation and time of sampling. International Journal of Animal Sciences. 6 : 122-127.

Jindal, S.K. and Ludri, R.S. (1990b). Growth hormone content in lactating crossbred cows and buffaloes. Asian Australasian Journal of Animal Sciences. 3: 319-322.

Jindal, S.K. and Ludri, R.S. (1990c). Insulin concentrations in lactating crossbred cows and buffaloes. International Journal of Animal Sciences. 5: 249-352.

Jindal, S.K. and Ludri, R.S. (1993). Body composition changes in crossbred cows and Murrah buffaloes during lactation. Asian Australasian Journal of Animal Science 6 : 577-580.

Jindal, S.K. and Ludri, R.S. (1994). Relationship between some circulating hormones, metabolites and milk yields in lactating crossbreds cows and buffaloes. Asian Australasian Journal of Animal Science 7 : 239-248.

Jindal, S.K. and Ludri, R.S. (1995). Certain metabolite concentration in lactating crossbred cows and Murrah buffaloes as affected by stage of lactation and diurnal variation. Buffalo Journal 11(3) 305-312.

Jindal, S.K. and M.L.Madan (1987) Future technologies for improvement in animal production system with special reference to Asia. Dairy Guide, Jan. 1987 pp 15-17.

Jindal, S.K. *et al.* (2009) Annual Report CIRG 2008-2009, pp 46, CIRG, Makhdoom, Farah, Mathura

Jindal, S.K.(1998). Buffalo semen freezing - protocols and fertility. Souvenir, Seminar on Problems of Buffalo reproduction and their solutions., CIRB, Hisar

Jindal, S.K., Mahendra Singh and R.S.Ludri (1989). Growth hormone for more milk production. Indian Dairyman 40(20): 82-84.

Jindal, S.K.and S.C.Chopra (1993). Buffalo semen freezing , Problems & perspectives. Indian Dairyman. 45 : 96-100.

Johnson, L. A., and R. N. Clarke(1988) Flow sorting of X and Y chromosome-bearing mammalian sperm: Activation and pronuclear development of sorted bull, boar, and ram sperm microinjected into hamster oocytes Gamete Research 21: 335 -- 343

Kamra, D.N. (2008) Rumen manipulation and livestock production-Lessons for the future. Lecture delivered to the participants of Winter School on " Recent Advances in Improvement of Productive and Reproductive Efficience of Goats through Physiological and Nutritional Interventions", at CIRG Makhdoom

Katska, L. And Smorag, Z. (1985) The influence of culture temperature on in vitro maturation of bovine oocytes. Anim. Reprod. Sci. 9: 205-212.

Keskintepe, L., G.M.Darwish, A.T.Kenimer and B.G.Brackett (1994) Term development of caprine embryos derived from immature oocytes in vitro. Theriogenology 42:527-535.

Kharche SD, Goel AK, Jindal SK, Sinha NK and Singh NP. (2008). *In vitro* Fertilization (IVF) Technology in Goats. Technical Bulletin No.34 CIRG Makhdoom Publication Page 1-18.

Kharche, S.D., A.K.Goel , Jindal, S.K.and Sinha, N.K. (2008) Birth of a female kid from in-vitro matured and fertilized caprine oocytes.. Indian Journal of Animal Sciences 78: 680-685.

Kilcawley, Kieran (2006) Enzyme technology for the dairy industry in Food Biotechnology, Vol 15-16, Kalidas Shetty (Ed) CRC Press, 1982 pp

Knox, D.P., Redmond, D.L.,Skerce, P.J. and New lands, G.F.J. (2001) The contribution of molecular biology to the development of vaccines against nematode and tramatode parasites of domestic animals. Vet. Para. 101: 311-315.

Kohler, G. and Milstein, C(1985) Continous culture offused cells secreting antibody of predefined specificity. Nature 256 : 495-497.

Kono, et al (2004) Birth of parthenogenetic mice that can develop to adulthood Nature 428, 860-864

Kumar, A., .A.Saini, S.K.Jindal & M.L.Punj (1998). Cryopreservation of embryos of various domestic animals. Indian Dairyman 50 (5) 5-10.

Kumar, A.,V.S.Solanki, G.C.Jain, S.K.Jindal and V.N.Tripathi (1997). Ultrasonographical studies of buffalo ovarian follicules during superovulation. International Journal of Animal Sciences 12 : 145-147.

Kundu, S.S., Kundu, S., Prasad, D.N. and Jakhmola, D.N. (1988) Lignin biodegradation to improve digestibility of straws. Int. J. Animal. Science. 3: 1-16.

Leibfried, M.L. and First, N.L. (1980) Effect of bovine and porcine follicular fluid and granulosa cells on maturation of oocytes in vitro. Biol. Reprod. 23: 699-704.

Lenz,S. And Lauritsen, J. (1982) Ultrasonically guided percutaneous aspiration of human follicles under local anesthesia: A new method for collecting oocytes for in vitro fertilization. Fertil. Steril. 38: 673-677.

Li, X.L., H. Chen and L.G. Ljungdahl, (1996). Cloning, sequencing and over-expression of a xylanase cDNA in Escherichia coli of the polycentric anaerobic fungus Orpinomyces sp. Strain PC-2. Proc. 96th Ann. Meet. Amer. Soc. Microbiol., New Orleans, pp: 361. USA.

Lightowlers, M.W. et al (2003). Vaccines against cestode parasites : anti helminth vaccine that work and why. Vet. Para. 115: 83-123

Linder, H.R., Bar-Ami, S. And Tsafriri, A. (1980). Gonadotropin induces meiosis in rat oocytes om vitro. In Animal Models in Human reproduction. Eds. Serio, M and Martini, L, Raven Press, New York pp 65-85.

Looney, C. R., B. R. Lindsey, C. L. Gonseth, and D. L. Johnson. (1994). Commercial aspects of oocyte retrieval and in vitro fertilization (IVF) for embryo production in problem cows. Theriogenology 41:67-72

Lopes S, Sun JG, Juriscova A, Meriano J, Casper RF. (1998) Sperm deoxyribonucleic acid fragmentation is increased in poor quality semen samples and correlates with failed fertilization in ICSI. Fertil Steril; 69:528 –32.

Lu, K.H. , Gordon, I., Boland, M.P. and Grosby, T.P. (1987a) In vitro fertilization of bovine oocytes matured in vitro. British Society of Animal Production, Winter meeting, 23-25th March, Scarborough, Paper No 28 p 2

Lu, K.H. , Gordon, I.,Gallagher, M. And McGovern, M. (1987b) Pregnancy established in cattle by transfer of embryos derived from in vitro fertilization of oocytes in vitro. Vet. Rec. 121: 259-260.

Madan, M.L. (1991) In vitro fertilization of buffalo oocytes. Symposium. Biotechnology in Animal Production, March 13-14, 1991, NDRI., Karnal, India

Madan, M.L. (1992) Embryo transfer technology in buffaloes. Indian Journal of Animal Reproduction 13:108-117.

Mahadaven. M.M., Trounson, A.O.. (1984). Relationship of tine structure of sperm head to fertility of frozen human semen. Fertil. Steril. 41, 287-293.

Mahi CA and Yanagimachi R (1976) Maturation and sperm penetration of canine ovarian oocytes in vitro. Journal of Experimental Zoology 196 :189–196

Manik R.S.; Singla S.K.; Palta P. (2003) Collection of oocytes through transvaginal ultrasound-guided aspiration of follicles in an Indian breed of cattle Animal Reproduction Science, 76: 155-161

Marks, J.L. and Ax, R.L. (1985). Relationship of non return rates of dairy bulls to binging affinity of heparin to sperm. J. Dairy. Sci., 68: 2078-2082.

Meeusen, Els N. T., John Walker, Andrew Peters, Paul-Pierre Pastoret, and Gregers Jungersen (2007). Current Status of Veterinary Vaccines Clin Microbiol Rev. 2007 July; 20(3): 489–510.

Minato, Y., and Toyoda, Y. (1982). Induction of cumulus expansion and maturation division of porcine oocytes cumulus complexes in vitro. Japanese J. Zootech. Sci. 53: 480-487.

Misra, A.K. (1993). Superovulation and embryo transfer in buffalo, progress, problems and future prospects in India Buffalo J. 9(1)13-24.

Mohanty, BB. (1988). Biotechnology in Animal Agriculture. Int. J. Anim.Sci. 3 : 101-109.

Molinia, F.C., Evans, G. and Maxwell, W.M.C. (1996). Fertility of ram spermatozoa pellet-frozen in zwitterion-buffered diluents. Reproduction, Nutrition, Development 36: 21-29.

Moreno-Lopez(1988). Nucleic acid hybridization techniques in the diagnosis of infectious diseases. Int. J. Anim. Sci. 3: 149-156.

Morgensterm, L. and Soupart, P. (1972). Oocyte recovery from human ovaries. Fertil. Steril. 23: 758.

Muller, E. and Wittkowski, G. (1986). Visualization of male and female characteristics of bovine fetuss by real time ultrasonics. Theriogenology 25: 571-574.

Murzamadiev, A.M., Dombrovskii, N.N., Isabekov, B.S. and Dzhienbaeva, R.S. (1983) Effect of serum obtained at different stages of the oestrous cycle on the maturation of oocytes in intact follicles. Seriya Biologicheskaya 4: 67-70 (ABA 1984, 52: 1828)

Mutter, L.R., Garden, A.P. and Olds, D. (1964) Successful non-surgical bovine embryo transfer A.I.Digest 12:3

Nesse, L.L. and Larsen, H.J. (1987) Lymphocyte antigens in Norwegian goats: Serological and genetic studies. Anim. Gen. 18:261-268.

Noland, T.D. and Olson, J.E. (1989) Calcium induced modification of the acrosome matrix in dignitonin - permeabilized guinea pig spermatozoa. Bio. Reprod. 40: 1057-1066.

Olar, T.T., Bowen. R.A.. Picket, B.W. (1989). Influences of extender, cryoprotectant and seminal processing procedures on postthaw motility of canine spermatozoa frozen in thaws. Theriogenology 3 1, 45 l-461,

Padha, Harish (1996). Biotechnology: Its global impact and relievance to India. Current Science 71 (9) 670-676.

Palmiter, R.D. et al (1982). Dramatic growth of mice that develop from eggs microinjected with methalothionein -growth hormone fusion genes. Nature (London) 300: 611-615

Parrish, J.J., Susko- Parrish, J.L., Weiner, M.A. et al (1988). Capacitation of bovine sperm by heparin. Biol Reprod, ,38: 1171 - 1180

Paul, S.S., D.N. Kamra , V.R.B. Sastry , N.P. Sahu and A. Kumar (2003) Effect of phenolic monomers on biomass and hydrolytic enzyme activities of an anaerobic fungus isolated from wild nil gai (Baselophus tragocamelus) Letters in Applied Microbiology 36 :377 – 381./

Paul, S.S., Kamra, D.N., Sastry, V.R.B., Sahu, N.P. and Agarwal, N. (2004) Effect of administration of an anaerobic gut fungus isolated from wild blue bull (Boselaplus tragocamelus) to buffaloes (Bubalis bubalis) on in vitro ruminal fermentation and digestion of nutrients. Animal Feed Science & Tech. 115:143-157

Pereira, R.M and C.C. Marques (2008). Animal oocyte and embryo cryopreservation Cell and Tissue Banking 9:267-277

Phillips, P.H. and Lardy, H.A. (1940). A yolk-buffer pabulum for the preservation of bull semen. *J. Dairy Sci.* 23:399–404.

Pierson, R.A. and Ginther, O.J. (1988), Ultrasonic imaging of the ovaries and uterus in cattle Theriogenology. 29: 3-2021-37

Pieterse, M.C., Kappen, K.A., Kruip, Th. A.M., Taverne, M.A.M. (1988) Aspiration of bovine oocytes during transvaginal ultrasound scanning of ovaries. Theriogenology 30, 751-762.

Pieterse, M.C., Vos, P.L.A.M., Kruip, Th. A.M., Willemse, A.H. and Taverne, M.A.M. (1991a) Characteristics of bovine oestrous cycles during repeated transvaginal, ultrasound-guided puncturing of follicles for ovum pick-up. Theriogenology 35, 401-413.

Pieterse, M.C., Vos, P.L.A.M., Kruip, Th. A.M., Wurth, Y.A., van Beneden, Th.H., Willemse, A.H. and Taverne, M.A.M. (1991b) Transvaginal ultrasound guided follicluair aspiration of bovine oocytes. Theriogenology 35, 19-24.

Polge, C., A. U. Smith, and A. S. Parkes. (1949). Revival of spermatozoa after vitrification and dehydration at low temperatures. Nature (Lond.) 164:666.

Prather RS, Sims MM, First NL(1989). Nuclear transplantation in early pig embryos. Biol Reprod; 41:414–418.

Purchase, H.G. (1986) Future applications of biotechnology in poultry. Avian Diseases 30 : 47-59.

Purdy, P.H. (2006). A review on goat sperm cryopreservation. Small Ruminant Research. 63 : 215-225.

Ravi Ranjan, Ramachandran, N., S.K.Jindal, N.K.Sinha, A.K.Goel, S.D.Kharche and A.K.S. Sikarwar (2009)Effect of egg yolk leves on keeping quality of Marwari buck semen at refrigeration temperature. Indian Journal of Animal Sciences 79(7)662-664.

Ritar, A.J., Salamon. S., (1983). Fertility of fresh and frozen-thawed semen of the Angora goat. Aust. J. Biol. Sci. 36, 49-59.

Roy, A. 1957. Egg-yolk coagulating enzyme in the semen and Cowper's gland of the goat. Nature (Lond.) 179:318.

Roy, A., Srivastava, R.K. and Pandey, M.D. (1956) Deep freezing of buffalo semen diluted and preserved in glycine-egg yolk medium. Indian J. Dairy Sci. 9:61

Ruff, G. and Lazary, S. (1985) Investigations on the goat leukocyte antigen (GLA) system. Anim. Genet. 16(Suppl. 1):123

Saake, R.G., (1972). Semen quality test and their relationship to fertility. NAAB Proceedings of the 4th Tech. Conf. on Anim. Reprod. Fertil., pp. 22-28.

Sahu, N.P., Kamra, D.N., and Paul, S.S. (2004) Effect of cellulose degrading bacteria isolated from wild and domestic ruminants on in vitro digestibility of feed and enzyme production. Asian Australasian J. Anim. Sci. 17(2)199-202.

Saiki RK, Chang CA, Levenson CH, Warren TC, Boehm CD, Kazazian HH Jr, Erlich HA. (1988) Diagnosis of sickle cell anemia and beta-thalassemia with enzymatically amplified DNA and nonradioactive allele-specific oligonucleotide probes. N Engl J Med ;319(9):537-41

Saiki RK, Scharf S, Faloona F, Mullis KB, Horn GT, Erlich HA, Arnheim N. (1985) Enzymatic amplification of beta-globin genomic sequences and restriction site analysis for diagnosis of sickle cell anemia. Science.;230(4732):1350-4.

Sakkas D, Urner FG, Bianchi PG, Bizarro D, Wagner L, Jaquenoud N, *et al.* (1996) Sperm chromatin anomalies can influence decondensation after intracytoplasmic sperm injection. Hum Reprod;11:837– 43.

Salamon. S., Ritar. A.J., (1982). Deep freezing of Angora goat semen: Effects of diluent composition. Method and rate of dilution on survival of spermatozoa. Aust. J. Biol. Sci. 35, 295-303.

Salisbury, G.W., Fuller, H.K. and Willett, E.L. (1941). Preservation of bovine spermatozoa in yolk-citrate diluent and field results from its use. *J. Dairy Sci.* 24:905–910.

Salisbury, G.W., N.L. VanDemark, and J.R. Lodge. (1978). Physiology of Reproduction and Artificial Insemination of Cattle. 2nd ed.W. H. Freeman Co., San Francisco.

Samper. J.C., Hellander, J.C.. Crabo, B.C., (1991). Relationship between the fertility of fresh and frozen stallion semen and semen quality. J. Reprod. Fertil. 44 (Suppl.), lo7- 114.

Santl, B, H.Wenigerkind, W.Schernthaner, J.Mödl, M.Stojkovic, K.Prelle, W.Holtz, G.Brem, E.Wolf (2009) Comparison of ultrasound-guided vs laparoscopic transvaginal ovum pick-up (OPU) in simmental heifers Theriogenology, 50 : 89-100.

Saxegaard, F and Fodstad , FH (1985)Control of paratuberculosis (Johne's disease) in goats by vaccinationThe Veterinary Record, Vol 116, Issue 16, 439-441

Selinger, L.B., C.W. Forsberg and K.J. Cheng, (1996). The rumen: a unique source of enzymes for enhancing livestock production. Anaerobe, 2: 263-284.

Shannon, P., and Vishwanath, R. (1995). The effect of optimal and suboptimal concentrations of sperm on the fertility of fresh and frozen bovine semen and a theoretical model to explain the fertility differences. *Anim. Reprod. Sci.* 39:1–10.

Sharma, M.C. and Mishra, R.R. (1987) Livestock Health and Management, Khanna Publishers, Delhi pp 1-591.

Sharma, M.C., Mahesh Kumar and Sharma, R.D.(2008) Textbook of clinical veterinary medicine. ICAR, New Delhi

Sharma, M.C., Pathak, N.N. and Lal, S.B. (2001) Liver : Structure, Disorders, Diagnosis and Therapeutic Management, IVRI, Izatnagar pp1-141.

Shen HM, Dai J, Chia SE, Lim A, Ong CN.(2002)Detection of apoptotic alterations in sperm in subfertile patients and their correlations withsperm quality. Hum Reprod;17:1266 –73.

Singh S.V., Singh, A.V., Singh, P.K., Sohal, J.S., Gupta, V.K. and Vihan, V.S. (2008). Therapeutic effect of a new"Indigenous Vaccine"developed using novel native"Bison Type"genotype of Mycobacterium avium subspecies paratuberculosis for the control of clinical Johne's disease in the naturally infected goat herds in India. Comparative Immunology Microbiology Infectious Disease. 24.09.07

Singh, Mahendra and Ludri , R.S. (1998) Immediate effect of bromocryptine on plasma hormone concentrations during early lactation in crossbred goats. Small Ruminant Research 31: 141-147.

Singh, Mahendra and Ludri , R.S. (1999) Plasma prolactin , blood metabolities and yield and composition of milk during early lactation in goats following administration of bromocryptine Asian Austr. J. Anim. Sci. 12:585-589.

Singh, Mahendra and Ludri , R.S. (2000) Plasma prolactin, blood metabolites and milk production in bromocriptine administered crossbred goats. Small Ruminant Research. 35: 255-262.

Slade, N.P.; Takeda, T.; Squires, E.L.; Elsden, R.P.; Seidel, G.E. Jr. (1985) A new procedure for the cryopreservation of equine embryos Theriogenology 24(1) p. 45-58

Southern, E.M. (1975) Detection of specific sequences among DNA fragments separated by gel electrophoresis. J. Mol . Biol . 98:503-517

Spooner, R.L. (1997) Genetics of disease resistance and the potential of genome mapping. Trop. Anim. Hlth. Prod. 29: 95S-97S.

Steptoe, P.C. and Edwards, R.G. (1978) Birth after reimplantation of human embryo Lancet 2:366.

Suresh, K.P., S. Nandi, S. Mondal (2009) Factors affecting laboratory production of buffalo embryos: A meta-analysis Theriogenology 72(7)978-985.

Techakumphu C, Lohachit P, Chantaraprateep P, Prateep P, Kobayashi G. Preliminary report on cryopreservation of Thai swamp buffalo embryos: manual and automatic methods. Buffalo Bull 1989;8:29–36.

Thibier M. (2003) More than half a million bovine embryos transferred in (2002): a report of the IETS Data Retrieval Committee. IETS Newsletter; 21(4): 12-19.

Thibier, M. and Wagner, H.G (2002) World statistics for artificial insemination in cattle. Livestock Production Science 74 : 203-212.

Tischner, M. (1987) Problems and perspectives of embryo transfer in horses. Farm Animal 249-5.

Toyoda, Y. and Chang, M.C.(1974) Fertilization of rat eggs in vitro by epididymal spermatozoa and the development of eggs following transfer J. Reprod. Fertil 36: 9-22.

Trinci, A.P.J., D.R. Davies, K. Gull, M.I. Lawrence, B.B. Nielsen, A. Rickers and M.K. Theodorou, (1994). Anaerobic fungi in herbivorous animals. Mycol. Res., 98: 129-152.

Turner, H.N. (1982). Origins of the CSIRO Booroola. In: Piper, L. R., Bindon, B. M. and Nethery, R. D. (Eds.), The Booroola Merino. CSIRO, Australia, pp. 1–7.

Utsumi, K., and Yuhara, M. (1974) Survival of frozen embryos : Freezing and storage of rat blastocysts in dry ice alcohol. Scientific Reports of the Faculty of Agriculture, Okeyama University, No 44: 24-27.

Valcticel. A.. De las Heras, M.A., Perez, L., Moses. D.F., Baldassarre, H., (1994). Fluorescent staining as a of the ejaculate, In: Hafez. E.S.E.(Ed.), Techniques of Human Andrology. Elsevier, Amsterdam.

Van Dam, R.H., D'Amaro, J., van Kooten, P.J.S., van der Donk, J.A. and Goodswaard, J. (1979) The histocompatibility comple GLA in the Goat Anim. Blood Gps. Biochem. Genet. 10: 121-124.

Vishwanath R. (2003);Artificial insemination: the state of the art. Theriogenology 59:571-84.

Wakayama, T. and Yanagimachi, R. (1998). Development of normal mice from oocytes injected with freeze-dried spermatozoa. *Nature Biotech.* 16: 639–641.

Wells, David N (2006) Cloning livestock. CAB Reviews: Perspectives in Agriculture, Veterinary Science, Nutrition and Natural Resources 1:039

Weng, Jing-Ke, Xu Li, Nicholas D Bonawitz and Clint Chapple (2008) Emerging strategies of lignin engineering and degradation for cellulosic biofuel production, Current Opinion in Biotechnology 19 : 166-172

Whittingham, D.G. (1968) Fertilization of mouse eggs *in vitro*. Nature, 220, 592–593.

Whittingham, D.G. (1971) Survival of mouse embryos after freezing and thawing. Nature : 233: 125-126.

Wideman, D., Dorn, C.G. and Kraemer, D.C. (1989) Sex determination of the bovine fetus using linear array real time ultrasonography. Theriogenology 31: 272.

Willadsen, S.M., Polge, C., Rowson, L.E.A. and Morris, R.M. (1976) Deep freezing of sheep embryos J. Reprod. Fertil. 46:151.

Willett, E. L., W. G. Black, L. E. Casida, W. H. Stone, and P. J. Buckner (1951) Successful Transplantation of a Fertilized Bovine Ovum Science 113. no. 2931, p. 247.

Wilmut, I. and Rowson, I.F.A. (1973) Experiments on low temperature preservation of cow embryos. Vet. Rec. 92:686-690.

Wilmut, I., Schnieke, A.E., McWhir, J., Kind, A.J. and Campbell, K.H.S. (1997) Viable offspring derived from fetal and adult mammalian cells. Nature 385: 810-813.

Wilson, J.D.(2000) Endocrinology : Survival as a discipline in the 21st century. Annu. Rev. Physiol. 62:947-950

Woolliams J. A. and Wilmut I.(1999) New advances in cloning and their potential impact on genetic variation in livestock Animal Science 68:245-256.

Yadav, P.S., Anil Saini, V.S.Solanki, O.K.Hooda and S.K.Jindal (1997). Comparison of follicular fluid and estrus serum on in vitro maturation ,fertilization and developlment of buffalo embryos. International Journal of Animal Sciences 12 : 193-196.

Yanagimachi R, Chang MC.(1963)Fertilization of hamster eggs in vitro. Nature; 200: 281-282

Yash Pal and Ludri, R.S. (1997) Circulating levels of some hormones and metabolities during initiation and early lactation in crossbred cows and buffaloes. Ph.D. Thesis, NDRI, Karnal.

Younis, A.I., Zuelka, K.A, Harper, K.M., Oliveira, M.A.L. and Brackett, B.G. (1991) In vitro fertilization of goat oocytes. Biology of Reproduction 44: 1177-1182

Zavos, P.M.. Correa. J.R., Sofikitis, N., Kofinas, G.D., Zarmakoupis, P.N.A. (1995). Method of short-term cryostorage and selection of viable sperm for use in the various assisted reproductive techniques. Tohoku J.Exp. Med. 176. 75-81.Dairy Sci. 71, 1638-1646.

Zhang, J. J., L. Z. Muzs, and M. S. Boyle. (1990). In vitro fertilization of horse follicular oocytes matured in vitro. Mol. Reprod. Dev. 26:361-365.

Zorn B, Virant-klun I, Meden-Vrtovec H. (2000) Semen granulocyte elastase: its relevance for the diagnosis and prognosis of silent genital tract inflammation. Hum Reprod 15:1978–84.

Extra References

Abhi, H.L. (1982) Note on the freezing of buffalo semen by Land Shut method and fertility of frozen semen in comparison to liquid semen under field conditions. Indian Journal of Animal Sciences 52:809.

Agrawal, K.P. and Bhattacharyya, N.K. (1983) In vitro preservation of goat embryos at 5°C Indian Vet. J. 53: 441-442.

Agrawala, P.L; Wagner, V.A. and Geldermann, H. (1992). Sex determination and milk protein genotyping of pre implantation stage bovine embryos using multiplex PCR. Theriogenology. 38 : 969-978.

Ahmad, M., Ahmad, K.M. and Khan, A. (1986) Cryopreservation of buffalo spermatozoa in tris (hydroxymethyl amino methane). Pakistan Vet. J. 6: 1-3.

Ahmadi A, Ng SC. (1999) Developmental capacity of damaged spermatozoa. Hum Reprod 14:2279–85.

Aitken RJ, Baker MA, Sawyer D. Oxidative stress in the male germ line and its role in the aetiology of male infertility and genetic disease. Reprod Biomed Online 2003;7:65–70.

Almodin, C.G., Moron, A.F., Kulay, L., Mingueti-Camara, V.C. Moraes, A.C., Torloni, M. R. A bovine protocol for training professionals in preimplantation genetic diagnosis using polymerase chain reaction. Fertility and Sterility. 84:895-899. 2005.

Anand, S.R. (1979) Water buffalo : Dilution and preservation of semen and artificial insemination. World Rev. Anim. Prod. 15: 69-79.

Bach rach, H. L. et al. (1985). Revue Sci . Tech. Office Int. Des Epizootics 2 : 629-653.

Bandhyopadhyay, S.K. and Roy, D.J. (1975) Freezing and fertilizing ability of spermatozoa Indian Journal of Animal Health 15(1) 37.

Bavister, B.D. , Leibfried, M.L. and Liebermann, G. (1983) Use of modified tyrode-lactate as fertilization medium for in vitro matured oocytes. Biol. Reprod. 28: 235-247.

Belak, S. and Ballagi-Pordany, A. (1993). Application of the polymerase chain reaction (PCR) in veterinary diagnostic virology. Veterinary Research Communication. 17 : 55-72.

Betteridge KJ (2003)A history of farm animal embryo transfer and some associated techniques. Anim Reprod Sci.; 79:203-244.

Bloomfield FH, Oliver MH, Hawkins P, Campbell M, Phillips DJ, Gluckman PD, Challis JR, Harding JE. (2003) A periconceptional nutritional origin for noninfectious preterm birth. Science.; 300:606.

Campbell, K.H.S. et al (1997) Sheep cloned by nuclear transfer from a cultured cell line. Nature 380:64-66.

Chauhan, M.S. and Anand, S.R. (1991) In vitro maturation and fertilization of goat oocyte. Indian J. Exp. Biol. 29(2) 105-110.

Chinnaiya, G.P. and Ganguli, N.C. (1980) Freezability of buffalo semen in different extenders. Zbl. Vet. Med. A 27: 563-568.

Choi, T.S., Mori, M. , Kohmoto, K. And Shoda, Y. (1987) Beneficial effect of serum on the fertilizability of mouse oocytes matured in vitro. J. Reprod. Fertil. 79: 565-568.

Church, R.B. (1989). Possibilities for genetic engineering of animals. Int.J. Anim. Sei. 4 : 1-6.

Crozet, N, Therm, M.C. and Chemineau, P (1987) Ultrastructure of in vitro fertilization in the goat. Gamete Research 18: 191-199.

Depypere, H.T., McLaughlin, K.J , Seamark, R.F. , Warnes, G.M. and Matthews, C.D. (1988) Comparison of zona cutting and zona drilling as techniques for assisted fertilization in mouse. J. Reprod. Fertil. 84: 205-211.

Dooley, V.D. (1984) Follicular oocytes maturation for use in bovine xenogenous and IVF Diss. Abstr. International 44: 3582.

Downs, S.M., Schroeder, A.C. and Eppig, J.J. (1986) Serum maintains fertilizability of mouse oocytes matured in vitro by preventing hardening of zona pellucida. Gamete Res. 15: 115-122.

Drost, Maarten (1991) Training manual for embryo transfer in water-buffaloes, FAO ANIMAL PRODUCTION AND HEALTH PAPER 84, FAO, Rome

Duran EH, Morshedi M, Taylor S, Oehninger S. Sperm DNA quality predicts intrauterine insemination outcome: a prospective cohort study. Hum Reprod 2002;17:3122– 8.

Eggen,A and Fries, R (1995). An integrated cytogenetic and meiotic map of the bovine genome. Animal Genetics. 26 : 215-236.

Erlich, H.A.; Gelfand, D. and Sninsky, J.J. (1991). Recent advances in the polymerase chain reaction Science. 252: 1643-1650.

Evenson DP, Jost LK, Marshall D, Zinaman MJ, Clegg E, Purvis K, et al. Utility of the sperm chromatin structure assay as a diagnostic and prognostic tool in the human fertility clinic. Hum Reprod 1999;14: 1039–49.

Farin PW, Crosier AE, Farin CE Influence of in vitro systems on embryo survival and fetal development in cattle. Theriogenology 2001; 55:151-170.

Galli C, Duchi R, Crotti G, Turini P, Ponderato N, Colleoni S, Lagutina I, Lazzari G. (2003)Bovine embryo technologies. Theriogenology; 59:599-616.

Georges, M. (1996). Biotechnology and animal production: Advantages and problems. Proc., 64th General Session, Office, International des Epizooties, Paris 20-24 May, W.H.O.

Goel, A.K. Kharche, S.D and Jindal, S. K. (2009) Efficacy of progestin implants-eCG to increase kidding rate in post-partum anestrus goats. Indian Journal of Animal Sciences 79:473-475.

Goel, A.K. Kharche, S.D, Jindal, S. K. and Sinha, N.K. (2005). Superovulatory response and embryo production potential of Jakhrana goats treated with porcine follicle stimulating hormone. Indian Journal of Animal Reproduction. 26 (1): 31- 33.

Goel, A.K. Kharche, S.D, Jindal, S. K. and Sinha, N.K. (2006). Embryo production and kids born through embryo transfer in Jakhrana goats treated with p FSH. Indian Journal of Animal Reproduction. 27 (1): 75 -76.

Gordon, J.W. and Talansky, B.E. (1986) Assisted fertilization by zona drilling : a mouse model for correction of oligospermia. J. Expl. Zool. 239: 347-354.

Green, D.P.L. (1987) Mammalian sperm cannot penetrate the zona pellucida solely by force. Expl. Cell Res., 169: 31-38.

Green, D.P.L. (1988) Sperm thrusts and the problem of penetration. Biol Rev. Camb. 63: 79-105.

Hanada, A. (1985) In vitro fertilization in goat. Jpn. J. Anim. Rproduction. 31: 21-26.

Handrow, R.R., First, N.L. and Parrish, J.J. (1969) Calcium requirement and increased association with bovine sperm during capacitation by heparin. J. Exp. Zool. 252: 174-182.

Hansen PJ, Block J . (2004)Towards an embryocentric world: the current and potential uses of embryo technologies in dairy production. Reprod Fert Dev.;16:1-14.

Hasler JF. (2003)The current status and future of commercial embryo transfer in cattle. Anim Reprod Sci.; 79:245-264.

Hasler, J.F. Cardey, E., Stokes, J.E., Bredbacka, P. (2002)Nonelecphoretic PCR sexing of bovine embryos in a commercial environment. Theriogenology. 58:1457-1469,.

Henkel R, Hajimohammad M, Stalf T, Hoogendijk C, Mehnert C, Menkveld R, et al (2004). Influence of deoxyribonucleic acid damage on fertilization and pregnancy. Fertil Steril;81:965–72.

Henkel R, Kierspel E, Hajimohammad M, Stalf T, Hoogendijk C, Mehnert C, et al. (2003)DNA fragmentation of spermatozoa and assisted reproduction technology. Reprod Biomed Online;7:477– 84.

Hespell, Robert B. (1987). Biotechnology and modifications of the rumen microbial ecosystem. Proceedings of the Nutrition Society, 46 , pp 407-413

Hughes CM, Lewis SE, McKelvey-Martin VJ, Thompson W. (1996) A comparison of baseline and induced DNA damage in human spermatozoa from fertile and infertile men using a modified comet assay. Mol Hum Reprod; 2: 613–20.

Ipomp,D, Good, 8.A., Geisert, R.D; Corbin, C.J. and Conley, A.J. (1995).Sex identification in mammals with polymerase chain reaction and its use to examine sex effects on diameter of day 10 or 11 pig embryos. Journal of Animal Sciences. 73 : 1408-1415.

Izaike, Y., Takahashi, S. S.Watanabe, S.Akagi, M. Yamaguchi and T.Tokunaga (2000)Recent advances in Bovine Biotechnology Asian Australasian Journal of Animal Sciences 13 (Supplement) : 247-250.

Jindal, S .K. Arvind Kumar, O.K.Hooda, V.S.Solanki & M.L.Punj (1998). Diagnostic ultrasound in reproductive physiology of domestic animals. Dairy Guide 25-26.

Jindal, S.K. (2001) Transgenic animals, Compendium of Summer School on Cryopreservation of germplasm for improving productivity of dairy buffaloes (1-21 June, 2001) Course Director, Dr.N.N.Pathak, CIRB , Hisar pp 87-88.

Jindal, S.K. and Jain, G.C. (1992) Daily sperm output of Murrah buffalo bulls based on depletion trials . Indian Vet. J. 69: 567-569.

Jindal, S.K., G.C.Jain & O.K.Hooda (1995). Effect of glycerol level on the freezability of buffalo semen. Indian Journal of Dairy Science 48(1) ; 71.

Johnson, L. A., and G. E. Seidel, Jr. (1999). Proc. Current Status of Sexing Mammalian Sperm. Theriogenology 52:1267–1484.

Johnson, L.A.; Flook. J.P. and Hawk, H.W. (1989). Sex preselection in rabbits: Live birth from X and Y sperm separated by DNA and cell sorting. BioI. Reprod. 41 : 199-203.

Kane MT .(2003)A review of in vitro gamete maturation and embryo culture and potential impact on future animal biotechnology. Anim Reprod Sci.; 79:171-190.

Katska, L., Kaufford, P., Smorag, Z., Duschinski, U., Torner, H. And Kamtz, W. (1989) Influence of hardening of the zona pellucida on IVF of bovine oocytes. Theriogenology. 32: 767-777.

Keefer, L.L. , Younis, A.L. and Brackett, B.G. (1990) Cleavage development of bovine oocytes fertilized by sperm injection. Mol. Reprod. Develop. 25: 281-285.

Keskintepe L, Pacholczyk G, Machnicka A, Norris K, Curuk MA, Khan I, Brackett BG . Bovine blastocyst development from oocytes injected with freeze-dried spermatozoa. Biol Reprod. 2002; 67:409-415.

Kharche SD, Goel AK, Jindal SK, Yadav EN, Yadav P and Sinha N K. (2009). Effect of serum albumin supplementation on in-vitro capacitation and fertilization of caprine oocytes. Small Ruminant Research. 81(2-3): 85 – 89.

Kharche, S.D., A.K.Goel , Jindal, S.K., Sinha, N.K. and Yadav, P (2008) Effect of somatic cells co-culture on cleavage and development of in-vitro fertilized caprine oocytes. Indian Journal of Animal Sciences 78: 686-692.

Kharche, S.D., E.N.Yadav, A.K.Goel, S.K.Jindal and N.K.Sinha (2008) Influence of culture media on in vitro fertilization of goat oocytes. Indian Journal of Animal sciences. 78 : 1075-1077.

Kharche, S.D., Goel, A.K., Jindal, S. K. and Sinha, N.K. (2006). In vitro maturation of caprine oocytes in different concentrations of estrous goat serum. Small Ruminant Research (The Netherlands) 64 (1 – 2):186 – 189.

Kharche, S.D., Goel, A.K., Jindal, S. K. and Sinha, N.K. (2006). In vitro maturation of caprine oocytes in different concentrations of estrous goat serum. Small Ruminant Research (The Netherlands) 64 (1 – 2):186 – 189.

Kubisch, H.M.; Hernandez, Ledezma, J.J.; Larsen, M.A.; Sikes, J.D. & Roberts, R.M. (1995). Expression of two transgenes in in vitro matrued and fertilized bovine zygotes after DNA micro injection. Journal of Reproduction and Fertility. 104 : 133-139.

Kubota, C. et al (2000) Six cloned calves produced from adult fibroblast cells after long term culture. PNAS 97: 990-995.

Kumar, A.,P.S.Yadav, A.Saini and S.K.Jindal (1997). Chronological studies of meiotic events during bubaline oocyte meturation in vitro. International Journal of Animal Science 12 : 197-200.

Kumar, A.,V.S.Solanki, S.K.Jindal, V.N.Tripathi and G.C.Jain (1997). Oocyte retrieval and histological studies of follicular population in buffalo ovaries. Animal Reproduction Science 47 ; 189-195.

Lee RSF, Peterson AJ, Donnison MJ, Ravelich S, Ledgard AM, Li N, Oliver JE, Miller AL, Tucker FC, Breier B, Wells DN Cloned cattle fetuses with the same nuclear genetics are more variable than contemporary half-siblings resulting from artificial insemination and exhibit fetal and placental growth deregulation even in the first trimester. Biol Reprod. 2004; 70:1-11.

Lopes, S., Jurisicova, A., Sun, JG.,Casper, RF.(1998).Reactive oxygen species :potential cause for DNA fragmentation in human spermatozoa. Hum. Reprod. 13:896-900.

Majumdar, A.C., Katiyar, P.K. , Raminder Singh, Taneja, V.K. and Bhat, P.N. (1989) Maturation of slaughter house ovarian follicular oocytes of buffalo in culture and subsequent in vitro fertilization Proc. World Buffalo Cong. New Delhi.

Manik, R.S., Madan, M.L. and Singla, S.K. (1994) Ovarian follicular dynamics in water buffalo superovulated in presence or absence of a dominant follicle. Theriogenology 41: 246.

Misra, A.K., Joshi, B.V., Agrawala, P.L.,Kasiraj, S.Sivaiah, N.S.Rangareddi and Siddique, M.U. (1992) Cryopreservation of bubaline embryos. Buffalo J. 3:297-303.

Niemann H, Kues WA. (2003) Application of transgenesis in livestock for agriculture and biomedicine. Anim Reprod Sci.; 79:191-201.

Ocampo, M.B., Ocampo, L.C., Rayos, A.A. and Hiroshi, Kanagawa (1989) Present status of embryo transfer in water buffaloes . Jpn. J. Vet. Res. 37: 167-180.

Parrish, J.J., Susko- Parrish, J.L., Weiner, M.A. et al (1988). Capacitation of bovine sperm by heparin. Biol Reprod, ,38: 1171 - 1180.

Pavithran, K., Vasanth, J.K., Rao, M. and Anantakrishnan, C.P. (1972) A method for deep freezing of buffalo semen. Indian Vet. J. 49:1125-1132.

Peeters, R.M. and Im, K.S. (1994) Trasgenic livestock- review. Asian Australasian Journal of Animal Sciences 7 : 1-17

Peterson AJ, Lee RS. (2003) Improving successful pregnancies after embryo transfer. Theriogenology; 59:687-697.

Pfeffer, M; Wiedmann, M. and Batt, C. A. (1995). Applica Hon of DNA Amplification techniques in veterinary diagnostics. Veterinary Research Communication. 19 : 375-407.

Pierson, R.A., Kastelic, J.P and Ginther, O.J. (1980) Basic principles and techniques in transrectal ultrasonography in horses and cattle . Theriogenology 29:3-20

Polejaeva, L.I.A. and Campbell, K.H.S. (2000) New advances in somatic cell nuclear transfer applications in transgenesis. Theriogenology 53:117-126.

Puha, A.C.Y., Abdullah, R. B., Mohamed, Z. (2003)A PCR based sex determination method for possible application in caprine gender selection by simultaneous amplification of the sry and ami-X genes. J Reprod and Develop. 49:307-3011,

Rao, K.B.C.A. and Totey, S.M. (1992). Sex determination in sheep and goats using bovine Y chromosome specific primers via polymerase chain reaction: potential for embryos sexing. Indian J. Experimental Biology. 30: 775-777.

Rasbech NO (1993). Artificial insemination. In: GJ King (Ed) Reproduction in Domesticated Animals. Elsevier, Amsterdam ,; pp. 365-386.

Rathore, A.K. (1965) Effect of instantaneous deep freezing (-79°C) on survival of buffalo spermatozoa with two levels of glycerol. Indian Vet. J.42:680.

Reddy, D.N.C., Gupta, D.C.L.R and Radhakrishan, T.D.G. (1982) Deep freezing of buffalo semen and its fertility studies. Indian Vet. J. 59: 574-575.

Robinson, J.J. and McEvoy, T.G. (1993). Biotechnology the possibilities Animal Production. 57: 335-352.

Robl, J. M., Prather, R. S., Branes, F., Eyestone, W., Northey, D., Gilligan, B. and First, N. L.: (1987).Nuclear transplantation in bovine embryos. J. Anim. Sci. 64:642-647,

Roy, D.J. (1974) A new field technique for deep freezing of bull and buffalo semen in "Tupol" Indian Vet. J. 51: 249-256.

Roy, D.J. and Ansari, M.R. (1973) Studies on deep freezing buffalo spermatozoa using two techniques Indian Journal of Animal Sciences 43: 1097.

Sahni, K.L. and Roy, A. (1972) Deep freezing of buffalo semen. Indian Vet. J. 49: 263-267.

Sahni, K.L. and Roy, A.(1972) A note on the effect of two storage temperatures on the keeping quality of sheep and goat semen in different diluents. Indian J. Anim. Sci. 42:580-

Sahni, K.L. and Roy, A.(1972) A study on the effect of deep freezing (-79*C) on post thawing revival of sheep and goat spermatozoa. Indian J. Anim. Sci. 42:102-105.

Sahni, K.L.(1968) Studies on quality, preservation and fertility of semen of sheep and goat. Ph.D. Thesis, Agra Univ. Agra.

Sakkas D, Mariethoz E, Manicardi G, Bizzaro D, Bianchi PG, BianchiU. (1999)Origin of DNA damage in ejaculated human spermatozoa. RevReprod;4:31–7.

Seidel GE Jr. (2003) Sexing mammalian sperm—intertwining of commerce, technology, and biology. Anim Reprod Sci.; 79:145-156.

Sergerie M, Laforest G, Boulanger K, Bissonnette F, Bleau G. (2005) Longitudinal study of sperm DNA fragmentation as measured by terminal uridine nick end-labelling assay. Hum Reprod;20:1921–7.

Sergerie M, Laforest G, Bujan L, Bissonnette F, Bleau G. (2005) Sperm DNA fragmentation: threshold value in male fertility. Hum Reprod; 20: 3446–51.

Shi L., Yue, W., Ren, Y., Lei, F., Zhao, J.: (2007)Sex determination in goat by amplification of HMG box using duplex PCR. Anim Repro Sci. 88,.

Sharma, M.C. and Mishra, R.R. (1987) Livestock Health and Management, Khanna Publishers, Delhi pp 1-591.

Sharma, M.C., Mahesh Kumar and Sharma, R.D.(2008) Textbook of clinical veterinary medicine. ICAR, New Delhi

Singh, Mahendra, Singh, R.P. and Singh, Omvir (2000) Artificial induction of lactation in cattle and buffaloes suffering from reproductive disorders. Indian Vet. J. 77: 237-239.

Singh, Pawan, S.K.Jindal, Sajjan Singh and O.K.Hooda (2000) Freezability of buffalo bull semen using different extenders Indian Journal of Animal Reproduction 21(1)41-42

Snyder, D.B. (1986) Improving animal health through monoclonal antibodies. Yearbook of Agriculture, Research for Tomorrow,94-98.

Stringfellow DA, Givens MD, Waldrop JG.(2004)Biosecurity issues associated with current and emerging embryo technologies. Reprod Fert Dev.; 16: 93-102.

Suzuki, Tatsuyuki, S.K.Singla, Sujata Jailkhani and M.L.Madan (1991) Cleavage capacity of water buffalo follicular oocytes classified by cumulus cells and fertilized in vitro. J. Vet. Med. Sci. 53: 475-478.

Totey, S.M., Singh, G., M.Taneja, C.H.Pawashe and Talwar, G.P. (1992) In vitro maturation , fertilization and development of follicular oocytes from buffalo (Bubalus bubalis) J. Reprod. Fertil. 95: 597-607.

Vajta G, Lewis IM, Trounson AO, Purup S, Maddox-Hyttel P, Schmidt M, Pedersen HG, Greve T, Callesen H. (2003)Handmade somatic cell cloning in cattle: analysis of factors contributing to high efficiency in vitro. Biol Reprod; 68:571-578.

Verma, M.C., Saxena, V.B., Tripathi, S.S., and Singh, R. (1975) A note on deep freezing of Murrah and Jersey bull semen. Indian Journal of Animal Sciences

Virro MR, Larson-Cook KL, Evenson DP. (2004). Sperm chromatin structure assay (SCSA) parameters are related to fertilization, blastocyst development, and ongoing pregnancy in in vitro fertilization and intracytoplasmic sperm injection cycles. Fertil Steril

Vlachos, K. and Tsakaloff, P.(1963) The artificial insemination of goats with deep frozen semen. Berl. Munch. Tierarztl. Wshnschr. 76:491 (ABA 32:2156).

Vlahov, K., Karaivanov, C., Petrov, M., Kacheva, D., Alexiev, A., Polihronov, O. & Danev, A. (1985). Studies on superovulation and embryo transfer in water buffaloes. Proc. First World Buffalo Congr., Cairo, Egypt. III:510–512.

Vlahov, K., Karaivanov, Ch., Petrov M. & Kacheva, D. (1985). Studies on superovulation and the production of embryos with the water buffalo (Bubalus bubalis) in Bulgeria. Vet. Med. Sci., 23:84–88. (In Bulgarian)

Wilmut, I; Haley, C.S. and Woolliams, J.A. (1992). Impact of biotechnology on animal breeding. Animal Reproduction Science. 28 : 149-162.

Wrathall AE, Simmons HA, Bowles DJ, Jones S. (2004). Biosecurity strategies for conserving valuable livestock genetic resources. Reprod Fert Dev. 16:103-112.

Yadav EN, Kharche S D, Goel A K, Jindal S K and Johri D.K. (2007). Comparative efficacy of different techniques for oocytes recovery from pre pubertal goat ovaries. Indian Journal of Animal Sciences 77 (10): 988 -990.

Yadav EN, Kharche SD, Goel A K, Jindal SK, Sinha NK and Johri DK. (2008). Effect of serum source on in vitro maturation and fertilization of prepubertal goat oocytes. Indian Journal of Animal Sciences 78 (2): 146 -149.

Yadav, P.S., Anil Saini , A.Kumar and G.C.Jain (1998) Effect of oviductal cell co-culture on cleavage and development of goat IVF embryos. Animal Reproduction Science 51:301-306

Yadav, P.S., O.K.Hooda, Anil Saini and S.K.Jindal (2001) In -vitro cleavage and development of buffalo oocytes with semen from different bulls. Indian Journal of Animal Sciences 71(8)752-754

Yadav, P.S., Razdan, M.N. and Dhanda, O.P. (1991) In vitro maturation of goat oocytes. Symposium, Biotechnology in Animal Reproduction, Karnal, India, Abstract 1-2.

Glossary

A	Adenine
AFLP	Amplified Fragment Length Polymorphism
AI	Artificial Insemination
AIDS	Aquired Immuno Deficiency Syndrome
Anaerobe	An organism that lives without oxygen. Eg. Rumen microbes
Anticodon	A specific sequence of three nucleotides in a transfer RNA, complementary to a codon for an amino acid in a messanger RNA.
Antigen	A molecule capable of eliciting synthesis of a specific antibody in vertebrates.
APC	Antigen-Presenting Cell
ART	Assisted Reproductive Technologies
ATP	Adenosine Triphosphate
AV	Artificial Vagina
BLAD	Bovine Leukocyte Adhesion Deficiency

BCG	Bacillus-Calmette-Guerin
bGH	Bovine Growth Hormone
BHK	Baby Hamster Kidney cells
BHV	Bovine Herpes Virus
BP	Base Pair, Two nucleotides that are in different nucleic acid chains and whose bases pair by hydrogen bonding eg A with T or U and G with C.Base Pair, DNA offers a means of storing and coding vast amounts of information captured by the sequence of bases present in the DNA strand. Human has about 3 x 10^9 base pairs in their genome.
bP	Base Pair
BSA	Bovine Serum Albumin
bST	Bovine Somatotropin
Buffer	Which helps in maintaining pH of a solution. A system capable of resisting changes in pH, consisting of a conjugate acid-base pair in which the ration of proton acceptor to proton donor is near unity. Eg Triss buffer, Phosphate buffer,
BVDV	Bovine Viral Diarrhoea Virus
CAMM	Computer Assisted Molecular Modelling
Carcinogen	A cancer causing chemical agent
CASA	Computer Assisted Semen Analyser
c-DNA	Cloned Deoxyribonucleic Acid, Complementary DNA
	A DNA that is usually made by reverse transcriptase and is complementary to a given messenger RNA: used in DNA cloning
CDRs	Complementarity-determining regions
Central Dogma	Genetic information flows from DNA to RNA to protein
Chimeric DNA	Recombinant DNA containing genes from two different species.
Chromatin	A filamentous complex of DNA, histones and other proteins constituting eukaryotic chromosomes.
Chromosome	A single large DNA molecule containing many genes and functioning to store and transmit genetic information.

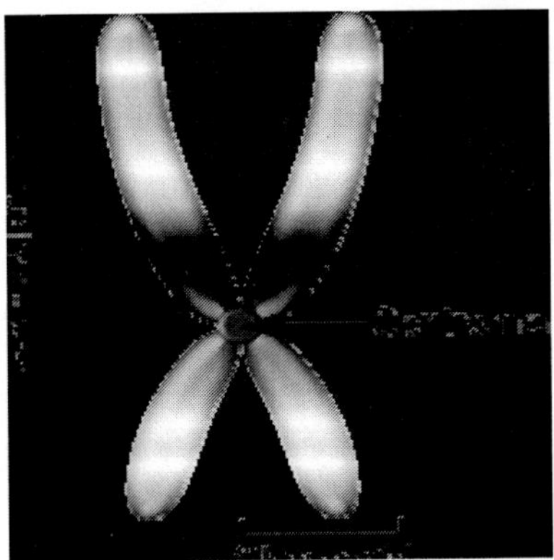

CIDR	Controlled Internal Drug Releasing Device
Clone	An exact replica of all or part of a macromolecule (e.g. DNA)
Codon	A sequence of three adjacent nucleotides in a nucleic acid that codes for a specific amino acid
CRT	Cathode Ray Tube
Cytoplasm	The portion of a cell's contents outside the nucleus.
Dalton	The weight of a single hydrogen atom
DMSO	Dimethy Sulphoxide
DNA	Deoxyribonucleic acid, A polynucleotide having a specific sequence of deoxyribonucleotide units and serving as the carrier of genetic information.Deoxyribonucleic acid

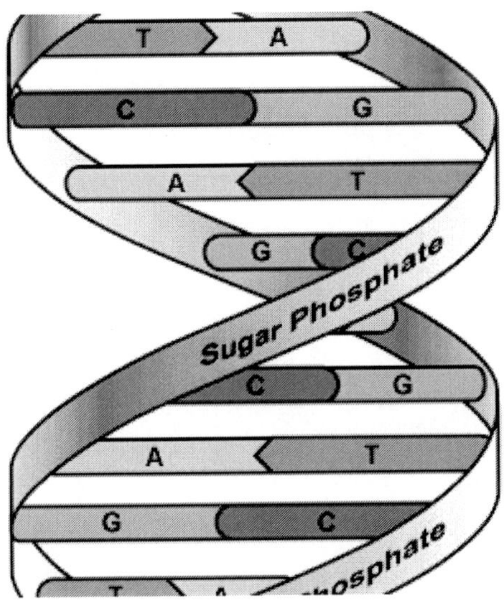

DNA Polymerase	DNA polymerase are specific enzymes that can synthesize a new DNA strand using DNA as template
DPT	Diphtheria, Pertussis and Tetanus
eCG	Equine Chorionic Gonadotrophin (a hormone having follicle stimulating properties.
EDTA	Ethylene Diamine Tetraacetic acid
EG	Embryonic Germ Cell
EGF	Epidermal Growth Factor
ELISA	Enzyme Linked Immuno Sorbent Assay
Enzyme	A protein specialized to catalyze a specific metabolic reaction
ET	Embryo Transfer
Exon	Protein-coding sequences
FCS	Fetal Calf Serum
FDA	Food and Drug Administration (US)-The FDA is responsible for protecting the public health by assuring the safety, efficacy, and security of human and veterinary drugs, biological products, medical devices, food supply, cosmetics, and products that emit radiation in the US.
FISH	Fluorescence *in situ* Hybridization – is a technique in which specific chromosomes can be made to fluoresce (glow) brightly under the microscope.
Fluorescence	Emission of light by excited molecules as they revert to the ground state.
FMD	Foot and Mouth Disease
FSH	Follicle Stimulating Hormone
FSH-O	Follicle Stimulating Hormone-Ovine
FSH-P	Follicle Stimulating Hormone – Porcine
G	Guanine
G-banding	Giemsa-banding – is a technique wherein chromosomes may be treated with various chemicals that produce alternating bands of light or dark intensity along the chromosomes.
Gene	A chromosomal segment that codes for a single polypeptide chain.

Genetic Code	The genetic code is the manner in which the nucleotide sequence in the mRNA specifies the amino acid sequence in proteins.
Genetic map	A diagram showing the relative sequence and position of specific genes along a chromosome molecule.
Genome	The genome is an organism's complete set of DNA components.
GH	Growth Hormone
GHG	Green House Gases
GIFT	Gamete Interfallopian Tube Transfer
GM	Genetically modified,
GnRH	Gonadotrophin Release Hormone
HIV	Human Immunodeficiency Virus
HLA	Human Leukocyte Antigen
hMG	Human menopausal gonadotrophin
Hormone	Hormone is a chemical substance that is secreted in trace amounts by an endocrine gland or tissue and acts as a messenger to regulate the function of another tissue or organ. Eg growth hormone, insulin, thyroid hormone.
HOST	Hyper Osmomolality Swelling Test
HPLC	High Pressure Liquid Chromatography
http	Hypertext Transfer Protocol (internet)
IFN	Interferons, A protein made by virus-infected cells of vertebrates: it prevents infection by a second kind of virus.Interferons
IGF-1	Insulin like Growth Factor-1
Immunoglo bulin	An antibody protein generated in response to specific antigen
Intron	Gene sequences which have no coding function
IPCC	Intergovernmental Panel on Climate Change
IPR	Intellectual Property Rights
IVF	*In vitro* Fertilization
IVM	*In vitro* Maturation
IVP	*In vitro* Production of embryos
JD	Johne's Disease

LH	Luteinizing Hormone
LH-RH	Luteinizing Release Hormone
MAb	Monoclonal Antibody
MHC	Major Histocompatibility Complex
Mitochondria	Membrane-surrounded organelles in the cytoplasm of eukaryotic cells: they contain the enzyme systems required in the citric acid cycle, electron transport and oxidative phosphorylation.
MOET	Multiple Ovulation Embryo Transfer
Molar solution	One mole dissolved in water and made upto 1000 ml.
mRNA	Messenger RNA, A class of RNA molecules, each of which is complementary to one strand of cell DNA and serves to carry the genetic message from the chromosomes to the ribosomes.
mt DNA	Mitochondrial DNA
NIH	National Institute of Health
NS	Normal Saline
Nucleus	In eukaryotes, a membrane enclosed organelle that contains chromosomes.
OCM	Oocyte Collection Medium
OEC	Oviduct Epithelial Cell
OIE	World Organization for Animal Health
Operon	An operon is a unit of prokaryotic gene expression. The operon includes the structural genes, the operator sequence and the regulatory gene(s).
OPU	Ovum Pick Up
PAGE	Poly Acrylamide Gel Electrophoresis
Pathogenic	Disease causing
PBS	Phosphate Buffered Saline
PCA	Per Chloric Acid
PCR	Polymerase Chain Reaction
$PGF_{2\alpha}$	Prostaglandin $F_{2\alpha}$
pH	Negative logarithm of hydrogen ion concentration of a solution. Indicates the degree of acidity or alkalinity of a solution.

Plasmid	A plasmid is an extra-chromosomal DNA molecule separate from the chromosomal DNA which is capable of replicating independently of the chromosomal DNA.
PMSG	Pregnant Mare Serum Gonadotrophin
PPR	Peste de Pestes Ruminantis
Prokayotes	The prokaryotes are a group of organisms that lack a cell nucleus, or any other membrane-bound organelles.

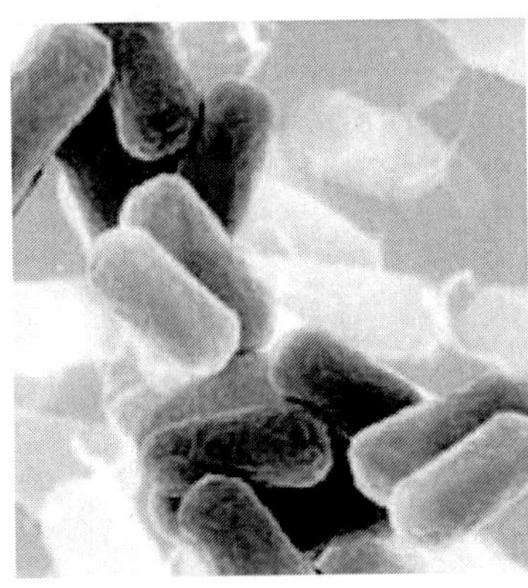

Promoter	The sequence of DNA required for RNA polymerase to bind to the template and accomplish the initiation reaction defines the promoter.
QTL	Quantitative Trait Loci
RAPD	Random Amplified Polymorphic DNA
r-DNA	Recombinant Deoxyribonucleic Acid – DNA formed by joining of genes into new combinations.
Recombinant antibodies	The new generation of monoclonal antibodies produced through recombinant DNA technology.
Replicon	The unit of DNA responsible for DNA replication is called replicon. It has an origin at which the replication is initiated and also has a terminus where the replication stops.
Retrovirus	RNA virus containing reverse transcriptase.
Reverse transcriptase	Reverse transcriptase is a type of polymerase which is involved in the replication of several kinds of RNA viruses. The reverse transcriptases are derived from avian myeloblastosis virus and Moloney murine leukemia virus. The ability of this enzyme to synthesize a DNA strand complementary to RNA template in the presence of suitable primer is central to cDNA technique.

Reverse transcription	In some retroviruses RNA can code for DNA or RNA can be reverse transcribed to DNA.
RFLP	Restriction Fragment Length Polymorphism
RIA	Radio Immunoassay, Sensitive quantitative determination of trace amounts of a hormone by its capacity to displace the radioactive form of the hormone from combination with its specific antibody. Radio immunoassay
RNA	Ribonucleic acid- a poly ribonucleotide of a specific sequence
ROS	Reactive Oxygen Species
RSV	Rous Sarcoma Virus
SCP	Single Cell Protein
SDS-PAGE	Sodium dodecyl sulphate Poly Acrylamide Gel Electrophoresis
SOET	Single Ovulation Embryo Transfer
Steroid hormones	A class of hormones from kidney and gonads which are lipids in nature containing the cyclopentanophenanthrene ring in their structure eg estradiol, testosterone
T CA	ThymidineTrichloro acetic acid
TALP	Tyrode's Albumin Lactate Pyruvate (TALP) culture medium
TCA cycle	Tricarboxylic acid cycle
TCM	Tissue Culture Medium
TMV	Tobacco Mosaic Virus
Transcription	Transcription is the synthesis of an RNA chain that is complimentary to one strand of DNA duplex and thus relays the information in the genes to the protein-synthesizing apparatus (Ribosome).
Translation	The genetic code specifies the means by which the four-letter nucleotide sequences in nucleic acids is translated into the entirely different amino acid sequence of proteins. Translation is a similar and much more specialized process in the cell.
tRNA	Transfer RNA- a class of RNA molecules which combines with a specific amino acid as the first step in protein synthesis.Transfer RNA
U	Uracil
USP	United States Pharmacopeia-The United States Pharmacopeia (USP) is an official public standards–setting authority for all prescription and over–the–counter medicines and other health related products.

Vaccine	A vaccine is a biological preparation that improves immunity to a particular disease.
Vector	Plasmids used in genetic engineering are called vectors.
Wobble	The relatively loose base-pairing between the base at the 3′ end of a codon and the complementary base at the 5′ end of the anticodon.
WTO	World Trade Organization
www	World Wide Web
ZIFT	Zygote Interfallpian transfer
ZP	Zona Pellucida

List of Nobel Prize Winners

List of Nobel Prize Winners in Physiology and Medicine

1901 E.A. von Behring (Germany)

1902 Sir Rohand Ross (England)

1903 N.R. Finsen (Denmark)

1904 Ivan P. Pavlov (Russia)

1905 Robert Kock (Germany)

1906 S. Romon Cajal (Spain) & Camillo Golgi (Italy)

1907 C.L.A. Laveran (France)

1908 Paul Ehrlich (Germany) & E. Metchnikoff (France)

1909 T-Kocher (Sweden)

1910 A. Kossel (Germany)

1911 A. Gullstrand (Sweden)

1912 Alexis Carrel (USA)

1913 Charles Richet (France)

1914 R. Barany (Austria)

1915-18	No Award
1919	J. Bordel (Belgium)
1920	August Krogh (Denmark)
1921	No Award
1922	A.V. Hill (England) and Mtto Meyerhoff (Germany)
1923	Frederic G. Banting and J.J.R. MacLeod (Canada)
1924	W. Einthoven (Holland)
1925	No award
1926	Johannes Fibiger (Denmark)
1927	J. Wagner-Jauregg (Austria)
1928	Charles Nocolle (France)
1929	Sir F.G. Hopkins (England) and C. Eijkman (Holland_
1930	Karl Landsteiner (USA) - Blood Groups
1931	Otto H. Warburg (Germany)
1932	Sir C.S. Sherrington and E.D. Adrian (England)
1933	T.H. Morgan (USA)
1934	G.R.Minto, W.P. Murphy and G.H. Whipple (USA)
1935	Hans Spemann (Germany)
1936	Sir Henry H. Dale (England) and Otto Loewi (Austria) - Chemical Control of heart rate.
1937	A. Szent Gyorgyi (Hungary)
1938	C. Heymans (Belgium)
1939	G. Domagk (Germany)
1940-42	No Award
1943	C.P. Henrik Dam (Denmark) and Edward A. Doisy (USA)
1944	Joseph Erlanger and Herbert Gasser (USA)
1945	Sir Alexander Fleming, Sir Howard W. Florey (England) and E.B. Chain (Germany)
1946	H.J. Muller (USA)
1947	Carl F. and Getty T. Cori (USA) and Bernardo A. Houssay (Argentina)
1948	Paul Mueller (USA)
1949	Walter R. Hess (Switzerland) and Antonia, C.A.F. Moniz (Portugal)

1950 Edward C. Kendall, Philip S. Hench (USA) and T. Reichstein (Switzerland)

1951 Max Theiler (USA - b, Africa)

1952 S.A. Waksman (USA)

1953 Hans A. Krebs (England) and F.A. Lipmann (USA)

1954 J.F. Enders, F.C. Robbins and T.H. Weller (USA)

1955 A.H.T. Theorell (Sweden)

1956 Andre F. Cournand, D.W. Richards (USA) and W. Frossmann (Germany)

1957 Daniel Bovel (Italy)

1958 G.W. Beadle, Joshua Lederberg and E.L. Tatum (USA)

1959 Severo Ochoa and Arthur Kornber (USA)

1960 Sir M. Burnet (Australia) and Peter B. Medawar (England)

1961 George von Bekesy (USA)

1962 Francis, H.C. Crick, Maurice, H.F. Wilkins (England) and James D. Watson (USA)

1963 Sir John C. Eccles (Australia), Andrew F. Huxley, A.L. Hodgkin (England)

1964 Konrad E. Bloch (USA) and Feodor Lyhen (West Germany)

1965 Francis Jacob, Andre Lwoff and Jacques Monod (France)

1966 Francis P. Rous and Charles B. Huggins (USA)

1967 Ranger Granit (Sweden), Haldon Keffer Hartline and George Wald (USA)

1968 Hargovind Khorana (USA), Robert W. Holley and Marshall W. Nirenberg (USA)

1969 Max Delbruck (USA), Alfred, D. Hershey (USA) and Salvador Luria (USA)

1970 Bernard Katz (England) Ulf von Euler (Sweden) Julis Axelrod (USA)

1971 Earl Wilbur Sutherland (USA)

1972 Gerald Edelman (USA), Rodney Porter (Britain)

1973 Karl von Frisch (West German), Zacharian Lorenz (Austria), Nicholas Tinbergen (Netherlands)

1974 Albert Claude (Luxembourg), George E. Palade (Hungary), Christian de Duve (Belcium)

1975	David Baltimore (USA), Renato Dulecco (Britain), Howard M. Temin (USA)
1976	Baruch S. Blumberg (USA), D. Carleton Gajdusek (USA)
1977	Rosalyn S. Yalow (USA), Roger C. Guillemin (USA) Andre V. Schally (USA)
1978	Werner Arber (Switzerland), Daniel Nathans (USA), Hamilton O. Smith (USA)
1979	Godfrey Hounsfield (Britain) Allan McCormack (USA)
1980	Buruf Benacerraf (USA) George Shell (USA) Jean Dausset (France)
1981	Roger Sperry (USA) David Hubel (USA) Torsten Wiesel (USA)
1982	S. Bergstroem (Sweden), B. Samuelson (Sweden) John Vane (Britain)
1983	Barbara McClintock (U.S.A).
1985	Michael S. Brown & Joseph Goldstein (USA)
1986.	Stanley Cohen and Rita Levi-Mondalcini (USA)
1987	Susumu Tonegawa (Japan)
1988.	James Black (UK), Gertrude Elion & George Hitchings (USA)
1989.	J.Michael Bishop, Harold E Varmus (USA)
1990.	Joseph E Murrey , E.Donnall Thomas (USA)
1991	Erwin Neher and Bert Saksmann (Germany)
1992	Edmond Fischer & Edwin Krebs (USA)
1993	Richard J Roberts (UK) Philip A. Sharp (USA)
1994	Alfred Gilman (USA) and Martin Rodbell (USA)
1995	Edward Lewis (USA), Christiane Nusslein Volhard (Germany) and Eric F. Wieschaus (USA)
1997	Stanley Prusiner(USA)
1998	Robert F Furchgott, Louis J Ignarro and Ferid Murad (USA)
1999	Günter Blobel,
2000	Arvid Carlsson, Paul Greengard and Eric Kandel
2001	Leland H. Hartwell, R. Timothy Hunt and Paul M. Nurse
2002	Sydney Brenner, H. Robert Horvitz and John E. Sulston
2003	Paul C. Lauterbur, and Sir Peter Mansfield
2004	Richard Axel, and Linda B Buck
2005	Barry J. Marshall, and J. Robin Warren

2006 Andrew Z. Fire, and Craig C. Mello

2007 Mario R. Capecchi, Sir Martin J. Evans, and Oliver Smithies for their discoveries of principles for introducing specific gene modifications in mice by the use of embryonic stem cells

2008 Harald Zur Hausen for his discovery of human papilloma viruses causing cervical cancer, Francoise Barre Sinoussi, and Luc Montagnier for their discovery of human immunodeficiency virus.

Index

A

Ai 18, 19, 20, 23, 24, 27, 30, 31, 96, 105
Antibody 99, 103, 111, 112, 113, 182
Antigen 58, 69, 70, 82, 90, 94, 96, 99, 102, 110, 113, 115, 172
Artificial breeding 23
Artificial insemination 14, 18, 19, 20, 22, 23, 26, 27, 29, 30, 144, 146, 171
Artificial vagina 24, 189, 199
Av 24

B

Bacteria 3, 84, 85, 86, 89, 90, 95, 102, 109, 117, 125, 182, 192
Bioinformatics 129, 130, 143, 177
Biotechnology 1, 2, 3, 5, 6, 7, 10, 11, 14, 17, 18, 31, 60, 62, 73, 82, 83, 86, 87, 88, 89, 90, 91, 92, 93, 100, 102, 107, 113, 115, 117, 118, 119, 120, 123, 126, 139, 140, 141, 143, 144, 145, 146, 147, 148, 151, 155, 157, 158, 159, 163, 165, 166, 167, 169, 170, 171, 172, 174, 175, 176, 177, 178, 179, 181, 182
Blastocyst 54, 70, 71, 73, 74
Bluetongue 100, 172
Bovine serum albumin 42, 193
Brucellosis 98, 110, 172

C

Canine distemper 110
Capacitation 36, 40, 41, 42
Casa 19
Cattle 3, 4, 7, 9, 11, 14, 18, 19, 22, 23, 28, 29, 31, 33, 34, 35, 36, 37, 40, 41, 44, 45, 47, 48, 52, 56, 57, 61, 76, 78, 79, 81, 82, 95, 99, 103, 104, 111, 116, 118, 125, 127, 129, 142, 144, 171, 172, 173, 189, 190

Cervix 29, 30, 48
Chromosome 58, 59, 60, 62, 92, 199
Cleavage 60, 70, 88, 117, 200
Cloning 4, 6, 14, 43, 53, 54, 55, 56, 57, 63, 66, 69, 76, 158, 163, 182
Corpus luteum 35, 49
Cryopreservation 20, 21, 22, 23, 25, 26, 27, 28, 32, 45, 46, 47, 71, 145, 166
Cryoprotectant 22, 45
Culture media 40, 41, 183, 192
Cytokines 118

D

DNA 2, 5, 12, 14, 19, 20, 26, 27, 28, 29, 53, 54, 55, 58, 59, 60, 61, 62, 63, 64, 65, 66, 67, 71, 72, 79, 80, 84, 85, 87, 88, 92, 93, 94, 96, 98, 99, 100, 101, 103, 104, 106, 108, 109, 116, 119, 126, 127, 129, 130, 153, 157, 159, 162, 163, 164, 165, 176, 177, 181, 182, 183, 184, 185, 186, 187, 189
DNA fingerprinting 176, 177
DNA polymerase 59, 186
DNA sequencing 187
Dog 37, 129
Domestic animals 2, 4, 5, 9, 13, 17, 20, 47, 50, 52, 66, 67, 68, 72, 73, 81, 83, 85, 94, 111, 125, 126, 129, 144, 145
Doppler ultrasound 50
Dot blot 67

E

Electroporation 85
Elisa 107, 110, 111, 113
Embryo 4, 6, 14, 21, 27, 32, 33, 34, 35, 36, 41, 43, 45, 46, 47, 49, 50, 52, 53, 54, 56, 57, 58, 59, 60, 69, 70, 76, 99, 105, 141, 142, 143, 144, 145, 166, 172, 191, 192, 194, 200
Embryo sexing 58, 59, 60
Embryo transfer 31, 32, 34, 35, 36, 43, 47, 50, 52
Equine chorionic gonadotrophin 33
Estradiol 77
Estrous cycle 39
Estrus goat serum 192
Extenders 22

F

Farm animals 6, 14, 18, 36, 37, 38, 50, 60, 61, 62, 73, 117, 119, 127, 141, 142, 157, 158, 163, 164, 178
Fertility 19, 20, 24, 25, 26, 27, 28, 30, 31, 36, 42, 43, 56, 77, 78, 136, 144, 195
Fertilization 3, 6, 28, 30, 36, 37, 38, 39, 41, 42, 43, 53, 68, 69, 70, 71, 73, 74, 77, 142, 166, 189, 195, 199, 200
Fetal calf serum 40, 192
Fetus 11, 73
Flow cytometry 19
FMD 95, 100, 103, 104, 111, 114, 172, 173, 182
Follicle 33, 44, 48, 49, 52, 53, 77, 196
Follicle stimulating hormone 33, 77
Follicular fluid 40, 48
Foot and mouth disease 93, 99, 103, 110, 111, 115, 157, 172, 173, 174, 178, 182

G

Gametes 14, 21, 22, 37, 45
Gel electrophoresis 110, 112, 189
Gene 2, 3, 4, 5, 14, 44, 57, 58, 61, 63, 65, 66, 67, 68, 70, 78, 79, 80, 83, 84, 86, 87, 88, 92, 93, 96, 98, 99, 101, 103, 117, 119, 120, 121, 122, 127, 129, 130, 141, 160, 161, 164, 165, 166, 167, 176, 182, 186, 187, 188, 189
Gene injection 44, 61, 67, 83
Genome 126, 127, 128, 129, 131
Genome 14, 27, 28, 61, 64, 65, 67, 68, 71, 78, 92, 96, 109, 116, 119, 121, 146, 160, 177
Genomics 2, 6, 13, 18, 72, 91, 115, 116, 119, 127, 128, 129, 130, 131, 172, 175, 177

Gestation 37, 47, 49, 50
Glycerol 19, 22, 25, 190
Goat 9, 10, 13, 14, 20, 22, 23, 24, 25, 27, 28, 29, 30, 31, 36, 37, 43, 44, 45, 49, 72, 78, 81, 104, 117, 143, 166, 172, 173, 192, 193, 195, 196, 197
Gonadotrophins 33, 39, 196
Graffian follicles 39
Growth hormone 11, 12, 18, 64, 65, 82, 83, 99, 117, 144

H

Hormone 3, 5, 11, 12, 18, 32, 33, 38, 44, 64, 65, 77, 81, 82, 83, 99, 117, 144
Horse 37, 45, 47, 78, 99
Host 14, 62, 64, 76, 84, 86, 94, 101, 108, 116, 121, 170, 182
Human 2, 3, 4, 6, 7, 8, 12, 13, 14, 17, 26, 27, 37, 38, 44, 45, 49, 50, 52, 57, 61, 62, 65, 68, 69, 72, 74, 75, 76, 77, 78, 92, 93, 94, 95, 96, 98, 99, 101, 104, 106, 111, 115, 116, 117, 118, 119, 120, 121, 122, 124, 125, 126, 127, 129, 154, 158, 159, 161, 162, 164, 165, 177, 178, 181
Human growth hormone 65, 99, 117
Hybrid 112, 113, 115
Hybridoma 172, 178

I

Immune system 94, 95, 96, 103, 106, 113
Immunoglobulins 113
Immunology 10, 64, 175, 176
In vitro fertilization 3, 6, 36, 37, 38, 41, 43, 69, 70, 74, 166, 199, 200
In vitro maturation 14, 36, 38, 39, 40, 43, 69, 197, 198, 199
Incubator 64, 118, 150, 151, 192, 197, 199

K

Killed vaccines 101

L

Lactation 11, 81, 82, 83, 171
Laminar flow 148, 149, 150, 151, 154
Ligases 182

M

Major histocompatibility complex (MHC) 72
Mare 36, 50
Mastitis 172
Micro injection 65
Milk 3, 5, 6, 8, 9, 10, 11, 12, 13, 14, 20, 23, 61, 62, 80, 81, 82, 83, 88, 90, 92, 102, 110, 117, 118, 125, 129, 144, 157, 163, 164, 171, 192
Moet 31, 32, 36, 37, 38
Morula 54, 70

N

Nucleus 27, 29, 31, 39, 54, 55, 56, 66, 70, 198

O

Oocytes 4, 26, 33, 36, 37, 38, 39, 40, 41, 43, 44, 46, 50, 52, 53, 57, 65, 66, 69, 70, 71, 195, 196, 197
Opu 3, 38, 52, 53
Ovulation 31, 36, 47, 79

P

Parasitic 93, 102
Parthenogenesis 14, 68, 69, 70, 71
Parturition 82
Phage 172
Plasmid 65, 85, 86, 106, 182
Pmsg 33, 115
Polymerase chain reaction 29, 59, 60, 107, 144, 153, 157, 186
Poultry 9, 91, 96, 102, 105, 116, 127, 171, 173
Pregnancy 4, 11, 27, 28, 30, 31, 44, 47, 48, 49, 50, 51, 56, 57, 69, 144, 172

Pregnancy diagnosis 47, 48, 49, 50, 51, 144
Proteomics 2, 91, 110, 112, 129, 130
Puberty 4, 52

Q

Quantitative trait loci (QTL) 129

R

RAPD 29
r-DNA 162, 163
Restriction fragment length polymorphism (RFLP) 59
RNA 67, 94, 100, 109, 119, 157, 182, 183, 185
Rumen 3, 4, 83, 84, 85, 86, 87, 88, 90, 125, 145

S

Semen 14, 18, 19, 20, 21, 22, 23, 24, 25, 26, 27, 28, 29, 30, 31, 33, 44, 78, 91, 166, 189, 190, 191, 192, 199
Seminal plasma 24, 26, 163
Sequence homology 33
Sexing 58, 59, 60, 91, 146
Sheep 3, 9, 14, 20, 29, 30, 31, 36, 40, 45, 47, 49, 54, 55, 56, 57, 60, 61, 62, 64, 65, 69, 72, 77, 78, 79, 80, 81, 102, 104, 110, 111, 117, 125, 127, 143, 144, 166, 171, 172, 173, 174
Sperm 25, 27, 40, 59, 194, 199
Superovulation 32, 35
Swine 103, 129, 165

T

Tissue culture 191
Transgenic technology 61

U

Ultrasonography 47, 48, 49, 50

V

Vaccine 77, 102, 104, 105, 112, 173, 177
Virus 94, 95, 96, 98, 99, 100, 101, 103, 104, 105, 106, 107, 108, 110, 111, 114, 121, 165, 172, 173, 178, 182

W

Water buffalo 9, 56, 78, 111
World wide web 141, 143

X

X Chromosome 58, 59

Y

Y Chromosome 58, 59, 60

Z

Zona pellucida 41, 198
Zygote 38, 62, 66, 70